GENETIC ENGINEERING

GENETIC ENGINEERING

Social and Ethical Consequences

Panel on Bioethical Concerns
National Council of the Churches of Christ/USA

(This is a report for study and comment, not an official policy statement of the NCCC/USA.)

THE PILGRIM PRESS
New York

Library of Congress Cataloging in Publication Data

National Council of Churches of Christ in the United
States of America. Panel on Bioethical Concerns.
Genetic engineering—social and ethical consequences.

Includes bibliographical references.
1. Genetic engineering—Moral and ethical aspects.
2. Genetic engineering—Social aspects. I. Title.
QH442.N35 **1984** 241′.6425 83-17320
ISBN 0-8298-0701-2 (pbk.)

The Pilgrim Press, 132 West 31 Street, New York, NY 10001

Contents

Preface

Each reader will find he or she has differences of opinion, perspective, and even disagreement about the issues included in and those omitted from this study report. Varied responses are to be expected for at least two reasons.

First, the full range and nature of the social and ethical questions raised by emerging technologies that alter genetic material are new. Specialists and the general public are just beginning to understand the current and potential consequences of genetic engineering. There is valid disagreement over the questions that ought to be considered serious—having immediate importance—and those that are speculative.

Second, the purpose of this study report is to encourage informed, responsible consideration of emerging social and ethical questions; it is *not* a final policy statement. It is designed to raise issues and to solicit responses from readers. It is a "work in progress."

My role as editor has been to adapt the original study report of the Panel on Bioethical Concerns of the National Council of the Churches of Christ/USA into a book that educates and engages the reader and provides a useful resource for study groups. This involved adding discussion questions and some brief suggestions for group leaders at the conclusion of each section. Section Two from *Splicing Life,* the President's Commission report was also edited and added. Finally, I edited and included the appendixes at the end of the book and designed the questionnaire. No substantive changes have been made to the text of the Panel's original study report, which was written by Rena Yocum.

The Panel on Bioethical Concerns of the NCCC/USA invites all readers to study its report and respond. Fill out the questionnaire (see pages 71–81) and return it to the Panel. In this report, the Panel has provided a helpful aid to begin to understanding some of the significant social and ethical questions raised by developments in genetic engineering.

Frank M. Harron
Editor

PANEL ON BIOETHICAL CONCERNS

NATIONAL COUNCIL OF THE CHURCHES OF CHRIST/USA

James Crumley, chairperson
Lutheran Church in America

Arleon Kelley, NCCC staff
Assistant General Secretary, NCCC

Rena Yocum, principal writer
The Village Church, Prairie Village, Kansas

J. Robert Nelson, writing team
Boston University, School of Theology

Karen Rahnasto, writing team
Lutheran Church in America

From the Governing Board, NCCC

Gabriel A. Abdelsayed
Coptic Orthodox Church in N.A.

Marguerite Bowden
United Presbyterian Church, USA

Cynthia Hale
Christian Church (Disciples)

C. J. Malloy, Jr.
Progressive National Baptist Convention

Leroy Nesbit
African Methodist Episcopal Church

Sung Kook Shin
United Presbyterian Church, USA

Elton O. Smith
The Episcopal Church

Robert Stephanopoulos
Greek Orthodox Archdiocese of N.&S.A.

Karen Werme
United Church of Christ

Charles J. Wissink
Reformed Church in America

From NCCC Units

Allan J. Bury

Thom Fassett

Rihito Kimura

Paul Sherry

Robin Smith

Doreen Tilghman

NCCC Staff

Kenyon Burke

Robert Parsonage

Jeffrey Gros

General Secretary of the NCCC

Claire Randall

• SECTION ONE •
The Issues

O Lord, our Lord,
How majestic is thy name in all the earth!

When I look at thy heavens,
  the work of thy fingers,
  the moon and the stars which thou hast established;
What are *human beings* that thou art
  mindful of them,
and their children that thou
  dost care for them?

Yet thou hast made them
  little less than a god,
and dost crown them with glory and *honor,*
Thou hast given humanity
  dominion over the works of thy hands . . .

—Psalm 8:1, 3–6

THE PRESENT CONTEXT

It is now possible for humanity to create, intentionally, new life forms that have never before appeared on this earth. By using the basic building blocks of life, the DNA molecules, new creations can replicate themselves and enter the mainstream of life.

It is also now possible to own the patent to the process by which certain life forms come into existence.

It is now possible to eradicate, intentionally, some genes considered "bad" and substitute "good" genes for them.

It is now possible to alter all life forms, intentionally, with a precision and speed never before known to humankind. It is also possible to alter life in such a way that it affects not only the present generation, but also the gene pool of all future generations as well.

Words that once were primarily the language of the church are now also the words of the current biological revolution. *Life, Death, Creation, New Life, New Day, New Earth* are now in the vocabularies of biological science, biotechnology and biobusiness.

Over eight thousand years ago the ancient Sumerians and Babylonians knew the ability of yeast to make beer. The Egyptians knew that yeast could make bread rise. For centuries, however, no one knew how yeast worked. In the seventeenth century, Dutch scientist Anton van Leeuwenhoek discovered microorganisms, tiny creatures in yeast never before seen or understood. Then in the nineteenth century, the work of Jean Baptiste Lamarck and Louis Pasteur made biology a recognized science and an established part of modern academia.

GLIMPSES

If progress in understanding the nature of life was slow in the past, it is now so rapid that many of us can hardly keep up with the new processes and discoveries, their blessings and their dangers. Just as a kaleidoscope changes before our eyes, we catch glimpses of life taking on new forms, new patterns, and new questions.

1. *Glimpse—risk.* A strand of a DNA molecule from one organism is spliced into the DNA molecule of another organism. Behold, a new life begins. Some, including some within the scientific community, believe that there is a potential danger to all humankind. Does this experiment, which requires such precision in process and yet is in itself imprecise in effect, contain a "Frankenstein factor"? How will such research be evaluated, monitored, and controlled?

2. *Glimpse—public policy.* In the past, the scientific community did its own monitoring and policing. Today, this community remains, to a large degree, self-regulating. If genetic engineering experiments are potentially dangerous to an unsus-

pecting, unknowing public, is there not a need for a public policy of review and regulation?

3. *Glimpse—human subjects.* The impoverished Third World has at times been the laboratory for experiments outlawed in wealthier countries. Unsuspecting people in Africa, Asia, and South America have been victimized by involuntary experimentation. Similarly, in America the poor, the ill, the elderly, ethnic minorities, prisoners, or the less educated have been used, without their awareness or consent, by a few unscrupulous doctors and researchers. Such experimentation and research can only be abhorred and condemned.

4. *Glimpse—control.* Genetic engineering assumes that someone has a plan for what or how genes ought to be used. Who is the engineer? What is the role of any one individual in decisions influencing his or her own life or the life of all posterity?

5. *Glimpse—public funding.* Scientists sometimes use knowledge gained by research financed by public funds in business endeavors as biotechnology enters the marketplace, yet the taxpayers do not participate in the profits. In biobusiness, is profit for a few justified in light of the original source of funding for the development of their discoveries?

6. *Glimpse—freedom of research.* Public sources of funding for research are drying up. Government loans and grants are disappearing. Industry is filling the void. In such circumstances, the one who gives the grants controls the area of research. Therefore, research is not open-ended or for the sake of new knowledge only; it is directed toward those areas that have the highest chance of paying back the investment. What are the long-term results? Who will evaluate the effects as free research in public universities and laboratories is increasingly paid for by private monies and subject to private control?

7. *Glimpse—patents.* In 1980 the Supreme Court of the United States (in *Diamond v. Chakrabarty*) authorized the patenting of life forms. The continuum from organic to inorganic is

very complex and exactly where to draw the line between the two is difficult. For centuries, industrial biology and microbiology have made use of live organisms for certain processes, such as fermentation and brewing. As we move up the continuum toward human use of medicines and drugs (e.g., insulin and interferon), the complication becomes intense.

Patenting grants exclusive possession and/or control of certain property, inventions, and/or processes. Should any individual, company, or school hold exclusive rights to a life form? What are the implications as technology grows and becomes even more sophisticated? Can or should this policy trend be reversed?

8. *Glimpse—access to information.* There is a growing aura of scientific secrecy in university communities that threatens the normal flow of information surrounding new experiments and discoveries. Heretofore new findings have been quickly shared in learned journals and the new knowledge has become public property. Today the possibility of a patent or the chance for a profit is stifling such sharing. This veil of secrecy can be in conflict with the public good. Who is monitoring this new phenomenon? Who can change it?

9. *Glimpse—distributive justice.* There are parts of the world in which malnutrition is common and health care, hospitals, doctors, and money are so scarce that diseases are rampant. The needs of people in such lands are enormous; and despite their own sometimes heroic measures, millions of lives of Third World citizens are lost.

At the same time, in the wealthiest sections of the First World, research on sex selection, cosmetic surgery, and other less essential matters is being conducted. Most of the benefits from current medical research are available to only a small fraction of the world's population. Who determines the priorities? What is a just distribution of medical research and services?

10. *Glimpse—suffering.* Americans are perhaps the most pain-conscious people in the world, and in the campaign to eliminate pain we have become the least able to deal with it. If

this is so, is the elimination of all suffering still our ultimate goal in medical research? Do people who suffer less contribute more to life? What will be the future role of genetic counseling and genetic modification?

11. *Glimpse—human gene pool.* Medical discoveries now make it possible to keep alive people who previously would have died at an early age. Such folk now pass along their weak genes to their posterity. The result, according to scientists, is that the total gene pool is weakened (e.g., there has been in recent years a 6% increase in the diabetic population and a marked increase in those needing visual assistance through glasses, etc.). Will we use genetic engineering to raise a more perfect race—and in the process increase prejudice toward the imperfect?

12. *Glimpse—plant and animal gene pool.* A living species can be wiped off the face of the earth. When a strain of corn that produced a high yield of grain but lost its resistance to disease was developed through hybridization, huge losses were experienced in several states. What if farms in all states had used the same hybrid? To what extent must we preserve the natural gene pool of plants? Of animals? Of *Homo sapiens?*

13. *Glimpse—interrelatedness of life forms.* We are once more brought to the stark realization that we as humans are related to and dependent on the whole of the world. "It's not nice to fool Mother Nature" is more than a phrase in a commercial; it reveals a profound truth. The use of toxic chemicals in farming in the Midwest can create havoc in the lobster beds of the Gulf of Mexico. A volcano erupts in Latin America and the consequences are felt by people in Russia. How we grow our plants and what we feed our animals affects the human body and perhaps our personalities as well. What regulation is appropriate to ensure that we do not destroy a valuable life form, damage a society, or harm a community because of a lack of foresight?

O Lord, our Lord,
What are human beings that thou art mindful of them . . . ?

Yet you have given humanity
dominion over the works of thy hands . . .

—Psalm 8

QUESTIONS TO CONSIDER

1. Of the thirteen "glimpses" described in this section, which three concern you more than the others? Why?
2. Can you recall some examples from the past when new knowledge clashed with existing values? (Some examples might be Galileo's trial; resistance to Darwin's theory of evolution; the political debate over the development of nuclear technologies.) What similarities and differences do you see between these precedents and emerging developments in genetic engineering? What roles have religious institutions and persons played in previous, similar situations.
3. What values and rights are at stake when public policy is lax or absent, especially when
   - the innovations are as far-reaching as the manipulation of genetic material, the "building blocks" of life?
   - there are unknown and even unknowable risks and potential abuses?
   - the profit motive emerges as a significant factor?
   - secrecy and competition replace scrutiny of scientists' work by other scientists and the public?
   - there are potential medical applications far beyond the traditional duties to relieve suffering or, when possible, to cure; specifically, to alter species and control evolution?
4. What values and rights are at stake when public policy attempts to regulate intellectual enquiry, research, and experimentation, especially when
   - it invades a scientist's laboratory, an academic institution, or private business?
   - the public's decision-making role for the scientific community is expanded?
   - private enterprise, in this case biotechnological enterprise, is regulated?
   - our technological progress is held back while other nations advance?
   - the potential medical benefits are great, but may be delayed?

5. What can Christians especially contribute to discussion of these issues? Cite biblical or theological statements if you wish, or describe Christian values and concepts you believe are especially relevant.

SUGGESTIONS FOR STUDY LEADERS

NOTE: These suggestions, which appear at the end of each section, are for the use of study leaders. They supplement your most helpful guide for leading group discussion of this study paper—Questions to Consider, provided above.

1. Read through the entire study paper, including the appendices, before anything else.

2. Two to four weeks before your study group begins, clip from the newspaper, and if possible have participants also cut out articles about developments in genetic engineering. These can be used to illustrate several important points: the frequency with which these issues arise even in the daily newspaper and the concrete ways in which they arise.

3. Match the clippings with the appropriate "glimpse" discussed in the text. At the discussion session, discuss how these newspaper stories illustrate the issues raised in the text.

• SECTION TWO •
Fundamental Facts

Many of the questions raised about genetic engineering cannot be explored without some understanding of the technical aspects of contemporary genetics and cell biology. Lack of information—or misinformation—not only provokes unwarranted fears but may even mean that legitimate and important questions remain unasked. Yet most Americans have had little formal training in biology, let alone in the specialized fields, such as micro- and molecular biology, that are involved in genetic engineering. Although a brief synopsis is plainly no substitute for a detailed education, some background may be helpful for nonspecialist readers. . . .

DISCOVERING LIFE'S MYSTERIES

What is remarkable about the science of gene splicing is not that it seems strange to laypeople—for all science is arcane to those who do not specialize in its study—but rather how unfamiliar it would be for the geneticists of even one generation ago. The existence of discrete inherited factors (later called genes) was postulated in 1865 by Gregor Mendel, a Moravian abbot who studied the patterns of inheritance in pea plants; his important work relied, however, on inferences about genes, not knowledge about their structure or functioning. Mendel's work

---

Because some understanding of genetic processes is necessary, we provide in this section a brief introductory discussion for the nonspecialist from *Splicing Life* (President's Commission for the Study of Ethical Problems in Medicine and Biomedical and Behavioral Research, U.S. Government Printing Office, Washington, DC, 1982, pp. 25–49).

lay forgotten until the beginning of this century, when the techniques of classical genetics were developed and physicians began to apply genetic knowledge in diagnosing conditions and in advising people about the conditions known to follow Mendelian patterns. Fifty years passed before Francis Crick and James Watson proposed the double helix as the structure for deoxyribonucleic acid (DNA), which is sometimes called the "master molecule of life" since almost all living things—including plants, animals, and bacteria—possess it. And the basic technique of gene splicing—a method for cutting and reuniting DNA—is itself only a decade old.[1]

Equally remarkable is that many new discoveries point to further unanswered—and perhaps even unanticipated—questions. The humbling reality of human ignorance is as relevant for those in industry and government who sponsor and regulate scientific research as it is for those who engage in that research. Any attempt to unravel more of life's mysteries can lead in unexpected directions, with unknown risks and benefits. The choices made about proceeding in one direction rather than another—or whether to proceed at all—are not simply matters of original scientific insight or intuition nor even of taking the "next logical scientific step." They also depend upon the judgment of individual scientists, laboratory directors, and public and private sector sponsors, drawing on analogy and conjecture, educated by experience, and reflecting personal and institutional values.[2]

Cells and genes. The human body is made up of billions of cells. Each cell has a particular function—cells in the gastrointestinal tract produce enzymes that digest food, bone cells provide structural support, and so forth. In spite of their markedly varied functions, most cells share the same structural organization—they have a nucleus, where the genetic information is stored, and cytoplasm, where the specialized products of the cell are made.

It has been thought that all cells in an organism normally contain exactly the same genetic information, with the exception of the germ cells (sperm and eggs), which carry only half. This information is located on individual packets called chromosomes, which come in pairs, half derived from each parent. Every species of plant or animal has a characteristic number of chromosomes. Humans usually have 23 pairs, or a total of 46; the germ cells have 23 chromosomes, one from each pair, while the somatic cells (the rest of the cells in the body) contain a full set of chromosomes. . . .

Each chromosome includes a long thread of DNA, wrapped up in proteins. DNA is made up of chemicals called nucleotides, consisting of one small sugar molecule, one phosphate group, and one of four nitrogenous bases, which can be thought of as the four letters in the genetic alphabet (*A*, *G*, *T*, and *C*).* . . .

When a cell divides into two daughter (or progeny) cells—a process called replication—a complete and faithful copy of the genetic code stored on each chromosome is usually transmitted to each daughter cell. Each half of the zipper acts like a template for the creation of its zipper-mate by drawing to itself free nucleotides, which then line up according to the *A*–*T* and *G*–*C* pattern. . . .

Although all cells in an organism carry basically the same genetic material in their nuclei, the specialized nature of each cell derives from the fact that only a small portion of this genetic material (about 5–10%) is active in any cell. In the process of developing from a fertilized egg, each type of cell switches on certain genes and switches off all the others. When "liver genes" are active, for example, a cell behaves as a liver cell because the genes are directing the cytoplasm to make the products that allow the cell to perform a liver's functions, which would not be possible unless all the genes irrelevant to a liver cell, such as "muscle genes," were turned off. . . .

THE TECHNOLOGY OF GENE SPLICING

Gene splicing techniques have been understood by scientists for only a decade. During that time, they have been used primarily in microorganisms. Though experiments with higher animals indicated the possibility of using gene splicing for human therapy and diagnosis, numerous hurdles had to be crossed before such steps could be taken. Recent research has

---

*The four letters are from the name of the base in the nucleotide: *A* for adenine, *G* for guanine, *T* for thymine, and *C* for cytosine.

cleared some of those hurdles, and work is under way that may conquer the rest much sooner than was thought possible even two years ago. . . .

Recombinant DNA techniques. It was once thought that genetic material was very fixed in its location. Recent findings demonstrate that genetic recombination (the breaking and re-linking of different pieces of DNA) is more common between and within organisms—from viruses and bacteria to human beings—than scientists realized. In fact, genetic exchange is a mechanism that may, in evolutionary terms, account for the appearance of marked variations among individuals in a given species.[3]

If DNA replication were the only mechanism for the transfer of genetic information, except for rare instances of mutation each bacterium would always produce an exact copy. In fact, three general mechanisms of genetic exchange occur commonly in bacteria.[4] The first, termed transduction, occurs when the genetic material of a bacteriophage (a virus that infects bacteria) enters a bacterium and replicates; during this process some of the host cell's DNA may be incorporated into the virus, which carries this DNA along when it infects the next bacterium, into whose DNA the new material is sometimes then incorporated.

In a second process, called conjugation, bacterial DNA is transferred directly from one microorganism to another. Some bacteria possess plasmids, small loops of DNA separate from their own chromosome, that give the bacteria the ability to inject some of their DNA directly into another bacterium. And third, bacterial cells can also pick up bits of DNA from the surrounding environment; this is called transformation.

These mechanisms—naturally occurring forms of gene splicing—permit the exchange of genetic material among bacteria, which can have marked effects on the bacteria's survival. The rapid spreading of resistance to antibiotics, such as the penicillin-resistance in gonorrhea bacteria and in *Hemophilus influenzae* (the most frequent cause of children's bacterial meningitis), documents the occurrence of genetic transfers as well as their benefit, from a bacterial standpoint.

The basic processes underlying genetic engineering are

thus "natural" and not revolutionary. Indeed, it was the discovery that these processes were occurring that suggested to scientists the great possibilities and basic methods of gene splicing. What is new, however, is the ability of scientists to control the processes. Before the advent of this new technology, genetic exchanges were more or less random and occurred usually within the same species; now it is possible to hook together DNA from different species in a fashion designed by human beings.

The key to human manipulation of DNA came with the discovery, in the early 1970s, of restriction enzymes.* Each restriction enzyme, of which about 150 have so far been identified, makes it possible to cut DNA at the point where a particular nucleotide sequence occurs. The breaks, which are termed "nicks," occur in a staggered fashion on the two DNA strands rather than directly opposite each other. Once cut in this fashion, a DNA strand has "sticky ends"; the exposed ends are ready to "stick" to another fragment that has been cut by the same restriction enzyme. Once the pieces are "annealed" and any remaining gaps are ligated, the "recombinant DNA" strand will be reproduced when the DNA replicates.

Recombinant DNA studies have been performed primarily in laboratory strains of the bacterium *Escherichia coli*, which is normally present in the human intestine. This bacterium possesses only one small chromosome, but it may also contain several ring-shaped plasmids. Plasmids turn out to be useful vehicles (or vectors) by which a foreign gene can be introduced into the bacterium. A plasmid can be broken open with restriction enzymes, and DNA from another organism (for example, the gene for human insulin) can then be spliced into the plasmid. After being resealed into a circle, the hybrid plasmid can then be transferred back into the bacterium, which will carry out the instructions of the inserted DNA (in this case, to pro-

---

*These enzymes, which make it possible to cut DNA at predetermined places, exist as part of the defense system that bacteria use to respond to foreign DNA (from a virus, for example). Restriction enzymes cut the DNA of the invader into small pieces, while another substance protects the bacteria's own DNA from getting sliced.

duce human insulin) as if it were the cell's own DNA. In addition, since plasmids contain genes for their own replication independent of bacterial DNA replication, many copies of the hybrid plasmid will be present in each *E. coli* cell. The end result is a culture of *E. coli* containing many copies of the original insulin gene and capable of producing large amounts of insulin.

The process of isolating or selecting for a particular gene is commonly called cloning a gene. A clone is a group all of whose members are identical. Theoretically, this technology allows any gene from any species to be cloned, but at least two major steps must be taken to make use of this technology. First, it is quite easy to break apart the DNA of higher organisms and insert fragments randomly into plasmids—a so-called shotgun experiment—but identifying the genes on these randomly cloned pieces or selecting only those recombinant molecules containing a specific gene is much more difficult. Because scientists do not yet fully understand what controls gene regulation, inducing expression of the inserted genes has been a second major hurdle. Recently, scientists have been successful in getting a recombinant gene to function in multicell animals and, with the discovery of what are termed transposable elements, even in correcting a defect in some fruit flies' genes.[5] This development serves as a reminder that many technical barriers that loom large are rapidly overcome. Of course, new knowledge sometimes also reveals further, unanticipated technical difficulties to be overcome. . . .

CURING GENETIC DISORDERS

In the immediate future, the most important applications of gene splicing techniques for human health will probably be in the creation of products—hormones, enzymes, vaccines, and so forth—for human consumption and in the development of genetic screening. But in the long run, direct use of the technique in humans can be expected to have an impact that is much more significant in terms of changing people's health and developmental status, and more novel and far-reaching in conceptual and psychological terms. During 1982, the prospect of direct application of gene splicing to cure human genetic diseases moved forward by large steps, although formidable hurdles remain.

The simplest form of human gene splicing would be directed at single gene mutations, which are now known to cause more than 2000 human disorders.[6] Such a defect in just one gene—although each human cell has as many as 100,000 genes—can have tragic and even fatal consequences. Existing treatments of genetic diseases are all palliative rather than curative—that is, they are merely aimed at modifying the consequences of a defective gene. In contrast, gene splicing technology offers the possibility of correcting the defects themselves and thus curing at least some of these diseases. The effects of gene splicing might be limited to the somatic cells of the individual being treated or might, intentionally or otherwise, alter the germ cells, thereby creating a change in the genes that would be passed on to future generations.

Somatic cells. The basic method proposed for using gene splicing on human beings is termed "gene therapy." This is defined as the introduction of a normal functioning gene into a cell in which its defective counterpart is active. If the mutant gene is not removed but merely supplemented, the cells may continue to produce the defective product alongside the normal product generated by the newly added gene.

Even further in the future is a theoretical possibility, sometimes referred to as "gene surgery," in which not only would the normal gene be added but the defective gene itself would either be excised or its function suppressed, so that it would no longer send out a message for a defective product in competition with the message from the inserted "normal" gene.

The technology, which researchers are now attempting to develop, involves four steps: cloning the normal gene, introducing the cloned genes in a stable fashion into appropriate target cells by means of a vector, regulating the production of the gene product, and ensuring that no harm occurs to the host cells in the patient. Only the first step—cloning a normal counterpart of a defective gene—is a straightforward matter with current knowledge and technology.

Introducing copies of the normal gene specifically to a particular set of target cells can, in theory, be achieved. Gene therapy offers the greatest promise for those single-gene defects in which an identifiable product is expressed in a discrete subpopulation of cells. For example, sickle-cell anemia and beta-thalassemia (also called Cooley's anemia) both involve alterations in the hemoglobin gene that is expressed in an accessible subpopulation of cells (that is, bone marrow cells) that could be removed from the body for gene treatment and then returned to the patient. These two diseases have therefore been among the early objects of attention for researchers designing gene therapy techniques.[7]

In most other cases, it is not practical to remove the target cells (such as brain cells in people with Tay-Sachs disease) for gene repair. A far more promising approach takes advantage of the distinctive properties of different cells, the unique markers each type of cell has on its surface. Once the unique marker for particular cells has been identified, it may be possible to construct a special "package," carrying copies of the normal gene, that will home in on this marker and deliver the new genes exclusively to the cells where the defective gene is active.

Once in the cell, the normal gene may persist as an independent unit, like a plasmid, or may integrate itself randomly somewhere in the DNA. The principal problem is inducing the host cell to produce the proper amount of the desired product.[8] Lack of expression of the normal gene would prevent the "therapy" from being effective, whereas excess production could be deleterious or even fatal. Although transposable elements of the sort that permitted new genetic material to be inserted in a nonrandom fashion and properly expressed in the experiments with fruit flies have not yet been identified in human beings, a comparable set of DNA appears to exist in human beings.

A final worry is that introducing a new gene may disrupt the functioning of the existing cells. For example, were the new piece of DNA to be spliced in the middle of another gene, it could create a gene defect that is worse than the defect the gene therapy was intended to correct. . . .

Germ-line cells. Thus far, attempts at gene therapy have focused on treating a discrete population of patients' somatic cells. Some researchers believe that certain forms of gene therapy that have been considered, such as the use of a virus to carry the desired gene to the patient's cells, might also affect germinal cells. Furthermore, gene therapy could also be applied to fertilized human eggs (zygotes) in conjunction with *in vitro* fertilization techniques.* Whereas the effects of genetic therapy on somatic cells would be expected to be limited to the individual patient treated, DNA therapy of fertilized eggs would probably affect all cells—including the germ cells—of the developing embryo; assuming normal birth, development, and reproduction, the individual would then pass on the altered gene to his or her offspring according to Mendelian rules. Zygote therapy would thus involve an alteration of the genetic inheritance of future generations and a significant departure from standard medical therapy.

To date, genetic engineering experiments using zygotes have been conducted for academic rather than therapeutic reasons. Several laboratories are currently working on fertilized mouse eggs. In one experiment, mice developed from zygotes injected with the rabbit hemoglobin gene were reported to contain rabbit hemoglobin in their red blood cells.[9] The medical significance is obvious. In a case where both parents are carriers of a particular recessive disorder the risk of an affected child is one in four. But if the relevant normal gene could safely be introduced *in vitro* to a fertilized egg of that couple, the individual who resulted from the egg would not have the disease and none of his or her descendants would be at risk for that disease.

Zygote therapy differs significantly from gene therapy on somatic cells in several ways. First, from the standpoint of the individual it may be useful in the treatment of genetic diseases, like cystic fibrosis, that affect many tissues—lungs, pancreas, intestines, and sex organs—rather than a discrete, accessible subpopulation of cells. Successful treatment at a very early

---

*The approach would involve the following: (1) isolating and amplifying the desire gene by standard recombinant DNA techniques, (2) removing a mature ovum from a woman and fertilizing it *in vitro*, (3) injecting copies of the cloned gene into the fertilized egg (zygote) using microsurgical techniques, and (4) implanting the genetically altered zygote into the woman's uterus.

stage of development would confer "good" genes to all the organs of an afflicted individual. Second, from the societal standpoint, such therapy if ever practiced on a vast scale could potentially reduce the overall frequency in the population of genes that usually have deleterious consequences, such as the sickle-cell gene.

Although zygote therapy may hold great promise, it is also fraught with technical risks and uncertainties. First of all, the technique itself is largely unproven, even with laboratory animals. For example, the success rate of microinjecting genes into mouse embryos remains low. Increasing the amount of DNA injected into a zygote makes it more likely that a gene will be incorporated, but it also increases the mortality rate of embryos. Microinjection of DNA into zygotes is obviously not a benign procedure.

The second major technical drawback at present is that transferred genes integrate randomly in the genome. Depending on the site of integration and perhaps the physiological state of the embryo, some of the foreign genes may be expressed and others not. Thus far, in experiments with mice, genes are rarely expressed in a tissue-specific way.[10] Even then, expression of the microinjected foreign gene in somatic tissue has not resulted in stable inheritance of that expression,[11] which is essential if the purpose is to introduce a new trait permanently. The consequences of having the wrong tissues producing the products of inserted genes could be disastrous.[12]

Finally, as in gene therapy on somatic cells, introducing foreign DNA into the zygote may affect the regulation of the cell in some undetermined way. Embryological development depends on a precise set of genetic instructions; disruption of this process is therefore much more likely to have serious adverse consequences than a disruption of the regulatory mechanisms operating in a subset of somatic cells. Instead of being therapeutic, therapy on zygotes or on more-developed embryos might be teratogenic and increase the incidence of congenital abnormalities.

In addition to the technical uncertainties involved, genetic manipulation of embryos raises serious ethical concerns. Altering the human gene pool by eliminating "bad" traits is a form of eugenics, about which there is strong concern. In 1982, the

Council of Europe requested "explicit recognition in the European Human Rights Convention of the right to a genetic inheritance which has not been interfered with, except in accordance with certain principles which are recognized as being fully compatible with respect for human rights."[13] . . .

GENES OR GENIES?

Biotechnology has made rapid advances in the past decade and will most likely continue to be a rapidly unfolding field. The awesome power entailed in these developments can be likened to the genie being let out of the bottle. As one observer of the field has noted:

> Some thirty-five years ago physicists learned how to manipulate the forces in the nucleus of the atom, and the world has been struggling to cope with the results of that discovery ever since. The ability to penetrate the nucleus of the living cell, to rearrange and transplant the nucleic acids that constitute the genetic material of all forms of life, seems a more beneficient power but one that is likely to prove at least as profound in its consequences.[14]

Stopping any enterprise out of a fear of potential evil not only deprives humanity of the fruits of new findings but also stifles strong impulses for innovation and change. Nevertheless, the technological allure of gene splicing ought not to be allowed to blind society to the need for sober judgments, publicly arrived at, about whether there are instances in which the price of going ahead with an experiment or an innovation will be higher than that paid by stopping the work.[15]

QUESTIONS TO CONSIDER

1. Did it surprise you to learn that genetic engineering technologies are in effect humankind's capacity to manipulate genetic processes, which happen "naturally" but randomly?
2. Which of these statements is closer to your chief reaction to this section?
   A. The potential benefits of genetic engineering, such as elimination or repair of some of the two thousand known genetic disorders, is very exciting.

B. Potential harms of genetic engineering, intentional or unintentional, outweigh whatever benefits there may be.

3. Which of these two statements is closer to describing your reaction to learning about our emerging capacities to unravel life's mysteries by discovering the essentials of genetic processes and to manipulate those processes?

   A. They are God's gifts to extend our role as co-creator.

   B. They represent an arrogant, and therefore sinful, and dangerous usurpation of God's prerogative to create life and direct its evolution.

4. Of these three lines of genetic engineering research—"gene therapy," "gene surgery," and "zygote therapy" (review the differences between each as described in this section)—which would you approve or disapprove? Why? Why not?

SUGGESTIONS FOR STUDY LEADERS

1. If you have a biologist or someone competent who can serve as a resource person, ask him or her to read this section ahead of time and present its most important points in a brief discussion. Leave time for answers and questions.

2. Request a film from your near-by public or university library. It should explore modern genetics, especially recombinant DNA technologies or gene splicing or genetic engineering. There are many good films available. Some of the best are from the PBS television program NOVA, distributed by Time/Life, Eisenhower Drive, Paramus, NJ 07652. Ask your librarian's advice.

• SECTION THREE •
Theology, Ethics, and Science

A journey has begun and we are uncertain of its destination and ramifications. The work in human genetics may revolutionize life as we know it.

Faithful people have been called to complete particular journeys, even when the destination is unknown. It was true of Abraham. It was true of Moses. It has been true of generations since. Faith in God makes the pilgrimage possible.

Our concern is not the problem of religion versus science. Rather we seek understanding of the new gift, which a large body of microbiologists has presented to our society. This gift is a rapidly growing quantity of knowledge about the genetic basis of all life, combined with a developing number of techniques by which forms of life, including human beings, can be modified. Through this new bioethical pilgrimage, we, as Christians (whether scientists or not), want to grow in our understanding of life and responsibility to humanity.

It may well be that this new turning point in molecular science is as radical as Copernicus's discovery in the sixteenth century that the earth revolved around the sun and not vice versa. It may be as far-reaching as Darwin's theory of evolution. It may be as crucial to our understanding of reality as Einstein's theory of the nature of matter. With each discovery we have had to rethink our scheme of the universe and the role of humanity in it.

Some will be tempted to say that we have received two messages from two different disciplines which have no relationship to each other. A second temptation will be to say that we have two contradictory messages between which one must choose. A third alternative is to acknowledge that we have been helped by the results of scientific work, that we need not automatically fear the impact of science, and that we must express our value judgments in dialogue with those in the fields of science and research. We may not judge the technology to be intrinsically good or evil; rather, we must speak to the appropriate use of the new knowledge. We need not hesitate to suggest that within the broad context of God's relation to humanity there is dependable knowledge about human life and the worldly environment.

Theology is the intellectual effort to discern the essence of the faith, to express and clarify its meanings, and to interpret particularity of beliefs. These beliefs are derived from the teaching of the Bible, the developing tradition of the church, tested by the broadest experience of the people and refined by their reasoning.

Even as we begin this new dialogue with science, we reaffirm some of our Christian beliefs that bear on this subject: God is Creator of all life. Life is precious; therefore, we must speak in faith to the manipulation of life forms. Humans have a role in the creative process. Life that enhances humanity is good. The transforming power of God is known as God engages with people in the task of enhancing human lives.

When the issues of genetic modification and/or the creation of new life forms are set alongside our faith affirmations, certain ethical questions arise. Ethical positions are worked out by the dialectical give and take between the realms of theology and technology in an interaction of mutual respect. Each has an effect on the other; both disciplines are needed.

GOD AS CREATOR OF LIFE AND COSMOS

As Christians, we believe in God the Creator. Some may use these words as an expression and not really believe them. Others think of divine creativity as an attractive theory in which they believe, but it exerts no influence on their understanding of life and the world. For us, creativity is the main principle conditioning our view of life and the world as well as our estimate of science.

The doctrine of creation not only addresses the baffling

problem of origination of matter—the world, organic and inorganic materials, human life—but also goes beyond origination to deal with meanings and purposes. God alone is acknowledged in faith as the sole Originator of all that exists. God is also the cosmic Purposer, whose good and loving will gives direction to both natural and human history. The confession that God is "the Maker of all things visible and invisible" testifies to the unique sovereignty of God. "Playing God" is a cliché for the human presumption of taking over God's creative power.

Christian faith goes beyond the impersonal theory of a First Cause, however. Creation is both continuous and personal. The Originator is also the Sustainer and Preserver of the universe, working constantly through the orderly natural processes and also—very recently in terrestrial time—through human intelligence, activity, and productivity. The more we learn through scientific investigation of the universe, from the unimaginably remote galaxies to living molecules and subatomic structures of matter, the more marvelous seem the works of divine creation. The more we learn through science about the particular units of the created order, the more we realize that the earth and the cosmos must be seen synoptically, as a whole and integrated system. Thus, the research, findings, and theories of scientists deserve our full and serious attention. However, they neither annul, displace, nor validate the belief in divine creation. The world is. In faith, we believe God creates it. "How majestic—the work of thy fingers, moon and stars—yet you have made humanity. . ." (see Psalm 8).

In traditional Christian theology God, who is the Source of creation, has provided a special place in it for humankind. We humans have been endowed with the powers to observe and reflect on all phenomena of the world, relating them to our interests. Our minds are informed in part by the knowledge of God's revealed intention and purpose for us—for our personal lives, humankind in its entirety, and the whole marvelous cosmos.

A high testimony to the value of each created human life and of all humanity was, and remains, the act of Incarnation. This is one of the foundation stones of the Christian faith. The life that was first blessed by being created in the image of God was confirmed and ratified by the becoming-human of the eternal Word of God in Jesus the Christ. In Jesus Christ humankind is re-created and renewed. This rejuvenation supplies force for the Christian witness to the original goodness and value of human life. Life is the created gift of God; that conviction can be further enhanced in this world and made eternal by God's action in Christ.

For these reasons, each and all human life is to be held in high respect. Traditionally, then, Christian theology regards the effect on human life as the primary theological criterion for making ethical judgments about genetic science.

We know that we and all human beings should be responsible for unborn generations of humanity. The human gene pool—that is, the totality of genetic material available for reproduction—is in danger of corrupting its offspring through imprudent, excessively risky genetic modifications. We have already demonstrated practices, either by ignorance or recklessness, that endanger future generations. The process of human striving against chance and adversity is concurrent with the evolution of biological forms and the changing conditions of geology and ecology. Human activity, performed in freedom and subject to moral distortion, contributes to these changes, as do the actions of insects and animals. "All creatures great and small" thus share the total phenomenon of life, with its endless complexity, its tragic wastefulness, and its often dubious purposes. But the human species, with its singularly distinct identity, dignity, and role in the composition of the organic world, has a correspondingly unique responsibility.

Life is holy because God is holy. This faith stance does not discount the scientists' assessment of life on a physicochemical basis. Rather, it affirms the geneticist who also strives for the enhancement of life. And yet we must maintain the awareness that *life* is always the central issue when we speak of genetic engineering and recombinant-DNA technology. Such technology is never just the joining of "mere bits of matter."

Just as the thought of Copernicus revolutionized our understanding of the position of the earth in the universe, so the new vision of the interrelatedness and interdependence of all life forms may cause us to reassess the position of the human species. No longer can ethics be confined to humankind;

our duty and our interest must include all life, the biosphere as a whole.

THE HUMAN ROLE IN THE CREATIVE PROCESS

Human beings, as members of a particular species of mammalian primates, belong to this created world. Our morphological similarities to other animals are evident; so are our requirements for air, water, and food similar to the needs of many others. Humans belong to the earth. We are not in essence other-worldly.

Human freedom, creative intelligence, intellectual self-transcendence, and love are ultimately inexplicable, however profoundly we may theorize about them. We can attest, describe, and appraise them. We account for them by reference to the divine creative power, will, and purpose.

The relationship of the individual human being to God, and of all humanity to God, is a mystery more profound and more elusive of explanation than any phenomenon or operating activity in the universe. In biblical faith it is expressed concisely by the phrase, *in the image of God.*

It is in the context of the *imago Dei* that we speak of mystery, spirit, thought, and agape. It is also in this context that we speak of the value of all and each human life. There are persons outside the Christian faith who also value life and hold all life sacred. For Christians, it is the understanding that we are created in God's image that puts a high value on human life and utilizes it to judge the uses and abuses of biotechnologies. Human dignity upsets conventional equations of cost/benefit and risk/success ratios.

The references in Genesis and in Psalm 8 for human beings to have "dominion" over the rest of creation are not a license for use or misuse. Rather, dominion carries with it a concept of custody, of stewardship, of being responsible for, of caring for all of creation. Therefore, we are called to live in harmony with all of creation, including humankind, and to participate with the Creator in the fulfillment of creation.

The language of co-creation must be used with care, however. Human creativity begins in the free processes of thought by the minds of persons and extends itself through language, inventiveness, and art. These powers are the created gifts of God to us human creatures, who in turn employ them for the developing of cultures, civilizations, and all the techniques by which millions of individuals create and change things.

How are these many created and creative gifts used? That is the moral question which is set before all human action and which pervades the existence of every person and every society. Are they employed for the debasing and exploiting of persons, or rather for the enhancing and enriching of their lives? Are the gifts of art, craft, and scientific technology exercised in harmony with the natural environment and responsible care for it? Or for ruthless, fruitless destruction? The bioethical concerns of Christians for respecting and protecting human life are linked as well to the idea of eco-ethical stewardship.

The biblical teaching of human "dominion" over all the earth now can be extended beyond the planet. It exalts the idea that men and women are coming into the full exercise of their given powers of co-creation. Never before has the practical reality of the concept of "dominion" been so evident as in our suddenly launched eras of nuclear energy, cybernetics, space exploration, and microbiological manipulation. These simultaneously born eras follow on the still effectual era of industrial and electronic technology. Considered together with all past prowess and craft, the scientific powers available to the present generation show both the glories and dangers of human creativity.

In fact, it is the dangers of human creativity that necessitate that we impose limits on the glories of our powers. Because there are dangers, both known and theoretical, in the applied techniques of genetic science, we concur with all who counsel certain restraints. We cannot agree with those who assert that scientific inquiry and research should acknowledge no limits. All that can be known need not be known if in advance it clearly appears that the risks are inordinate. Remembering its own history, the church should not oppose scientific advance, but it must speak out in judgment when the quest for new knowledge supersedes all ethical concerns.

For example, the theory of eugenics implies that the human race can be improved in various ways—physical strength, resistance to disease, intelligence, even psychic and moral excellence. It also involves a common concept of who and what is superior, who and what is inferior. Because of cultural biases, it could militate against ethnic minorities, women, and, of course, all who are physically and mentally "abnormal" or "defective."

Research and technical development pave the way for such eugenic programs by expediting breeding that would otherwise require many generations. Such procedures would begin with genetic screening and counseling, artificial fertilization methods, and, for some, sterilization. Even if there were a "positive eugenics program," it would probably require social organizing and involuntary compliance.

Christian concern for personal integrity and value, therefore, makes involuntary eugenic research morally unacceptable.

THE HUMAN CONDITION

The church has taught that we are created in the image of God, endowed with prodigious intelligence, and commissioned to exercise careful dominion over all of nature, including ourselves. We, as human beings, might by now have produced an earthly paradise. We certainly have not, nor are we doing so now. Why?

The church also teaches us that there is, paradoxically, within each of us the will to do good and the inability to do it. We also have a disposition toward egoism and exclusive self-interest. Our ambition tempts us to build self-exalting towers of Babel. Sin is both pride and irresponsible exercise of power. Sin is human rejection of both the Creator's love and the divine laws of justice and responsibility. Sin is self-deception. Sin turns the most ingenious of human achievements into instruments of injustice and malfeasance. Sin can turn even the best of human acts and intentions into occasions for self-gratification and cause us to inflict senseless injury on others.

Sin permeates all of life as a part of the human condition. This permeation is true individually and corporately, whether we are speaking of the church or the scientific community. Understood as pride and arrogance, sin can be witnessed as unwarranted confidence in one's own wisdom. Many contemporary discussions include questions or warnings about "playing God." This issue has arisen because human beings now have the ability to do "God-like" things, to direct and redirect the life processes of nature. It is not a matter of intending evil, it is a matter of being human—mortal, finite, imperfect, and sinful. Sin is the risk required and the price paid for freedom. To be human is to enjoy all the risks and benefits implied for existence by freedom: free thought, free decision, free action. The exercise of creativity in the realms of science and social organization could not occur without this divinely conferred freedom, yet the same freedom produces the possibility for conditions under which people experience injustice and suffering.

Christians believe that God is never indifferent to suffering, that God is present in the midst of all human struggle and suffering. God participates in the struggles and suffering of persons because human freedom must be preserved. However, there are varieties of suffering: there is suffering that seems grave because we lack the courage or stamina to bear it; there is suffering accepted and endured out of respect or love for another person, for an idea, a cause or a faith. Among these varieties of suffering, some are clearly evil, some are beneficial, some are morally ambiguous.

One valid goal of the life sciences and genetics in particular is the reduction of suffering. Surely the relief of suffering, in a world where suffering is both intense and for many unending, is an act of mercy; it was clearly a part of the ministry of Christ. Yet that affirmation must not lead us to the conclusion that suffering is the absolute enemy of human dignity.

Certainly as Christians we cannot glorify some abstract ideal of health as if physical well-being is an end in itself. Decency, love, justice, compassion, peacefulness, and dignity cannot be genetically programmed, nor can the worth of an individual depend on the presence or absence of disease. We must change the opinion that all suffering is evil and that painlessness (hedonism) is always best. The very axis of the Christian faith is the perception that the suffering that accompanies sin can be overcome by the divine sharing in it—the cross.

The main thrust of the gospel of the cross appears to agree with the purpose of such applied scientific techniques as genetic manipulation. Both are directed against the kind of suffering that injures or destroys life and its goodness. Nevertheless, preoccupation with the quest for painlessness and the attainment of an abstract notion of perfect health is likely to heighten a person's disposition toward self-interest, and a self-serving attitude can inhibit one's giving of personal sympathy, energy, and time to the needs of others. It can also breed an attitude of condescension, discrimination, or contempt for those who are handicapped, disabled, or otherwise severely disadvantaged. We do not recognize pain and suffering as good in and of themselves. However, what appears to be a "good" goal, the elimination of all suffering, can be transmuted into a distortion of the value of life and living. The mystery of sin lies in the obvious truth that the best of intentions may lead to corrupting consequences. Suffering can become an occasion through which God's purpose is fulfilled. It can point the way to participatory suffering—the awareness that as one suffers, so do we all.

What, for instance, should be the just proportion between resources expended for genetic research and those expended to alleviate world hunger? Are we willing to face the global aspects of suffering as well as our own? The prime symbol for our faith continues to be the cross of Christ. Its message of divine suffering with and for all humanity gives us hope in light of and despite our human condition.

THE TRANSFORMING POWER OF GOD

Christian faith is incompletely described in terms of creation and the human condition; it is also reconciliation, redemption, and transformation. Christian trust in and loyalty to a faithful God who works with people everywhere in nature and in history is found in the life, death, and resurrection of Jesus Christ. This "new creation" of life has a new dimension of hope and confidence and joy.

As Christians, we believe that we are both creatures of God and participants in the creative process as well. The whole of life is a chain of which humankind is only a part. We cannot just by our determination make everything come out right. The relative or imperfect good of human persons and institutions is compounded by the partial knowledge we possess; in respect to almost anything it can be said "we know in part."

Christian faith acknowledges this limitation and still it encourages the quest for more knowledge. God is truth. The practical implication of this encouragement is that risk must be taken. Once our limitations are confessed, the risks acknowledged, then we can know that it is only the transforming power of God which releases us to do our best.

The familiar dualism of matter and spirit, body and soul, is an ancient Indian and Greek idea. It is contrary to the nature of human life as understood by the Jews and early Christians. In our Hebraic heritage, a human person is literally psychosomatic, an integrated unity of body and soul. This concept rests on the prior belief that everything material has been created by God and that all creation is good. Human bodies are called temples of the Holy Spirit by Paul, rather than prisons of the soul, as Plato said. Christian faith looks in hope, therefore, not to a naturally immortal soul, but to the resurrection of the body and the life of the world to come. Human existence in finite time and space, here and now, is significant for each one's renewed and risen life to come.

Grace is God's activity in all dimensions of life. Grace is a relationship with God that we have not earned and cannot earn. It therefore gives us a sense of possibility, of new beginnings. The role of the scientists is also redemptive if and when it seeks human wholeness. This is not to say that human action is identical with God's initiative in human life, only that the goal is the same kind of goal, the wholeness of life.

Because of this high evaluation of mortal life in its wholeness, we in the Christian faith are committed to assisting one another throughout human society to enjoy healthful and wholesome conditions of living. As a Christian Church we want to witness to the grace we have received by serving others. The church has a special role in providing mutual support, growth, and fulfillment. Having received the gift of life from God, we in turn want to participate in gracious activity by seeking wholeness rather than brokenness, freedom rather than oppression, healing rather than suffering. The purpose of our individual and community life (whatever our vocation) is to respond to that Creating and Sustaining Love and to express it in real and tangible ways to all people. Therefore, scientific techniques that can affect both the organic substance of human beings and their environment are matters of concern to Christians. Indeed, the New Testament links the renewal of life for human beings to the transformation of the whole of God's creation.

Because all human life is lived in mutuality of community, within which each person is of incalculable value, it is wrong to contend that the rights of our society must always take precedence over those of the individual, or vice versa. The integrity and dignity of one human being cannot be sacrificed for the anticipated benefits of the society, especially not for the benefit or profit of the few. We see in the development of genetic technologies the inherent danger of persons being exploited and deprived of common resources. People may be put at risk involuntarily in the name of the common good. People who are likely to be affected must be protected from paternalistic attitudes of the scientific and technological elite.

Nevertheless, human inventiveness may enhance the lives of individuals and of societies. When it does, our appropriate response should be one of thanksgiving to God, of doxology. God's continuing creation takes place in history, and the history of humanity today is one of scientific exploration.

Life is of God. Through this transforming power of God, our human efforts and imperfections can contribute toward the quest for wholeness. The church is called to witness to this enabling grace of God. The purpose of our individual and community life as the church is to respond to God's creating and sustaining love and express it in real and tangible ways to all people. Therefore, we must express our values and relate them to this particular issue in history.

We are only human. However great the discoveries, however tantalizing the options, however grandiose the scheme, still we are human. We are subject to error, failure, and sin. When the subject is recombinant DNA, when the concern is the total gene pool for humankind, when the scenario is extinction of species or eugenic programs, or when the whole biosphere can be affected by any one experiment with new life forms *ad nova*, extreme caution can be our legitimate response.

Nevertheless, it is exciting to gain new knowledge, to break new barriers. It is exciting to observe biotechnology unraveling the key to genetic diseases. We celebrate the healing possibilities that are now before us through this new life-technology. Jesus said, "I came that they might have life and have it abundantly." When we have this same goal (whether as scientists or not), we are partners with one another in the ongoing journey for humankind; we are partners in the continuing creative process with God.

We have begun our journey. We are uncertain of its destination and ramifications. The agenda will change. Work in genetics may revolutionize life as we now know it. Faith in God will continue to make the pilgrimage possible.

QUESTIONS TO CONSIDER

1. Assuming that new knowledge and technologies in genetic engineering are God's gifts, for what purposes do you think they are intended to be applied:
   - to reduce suffering?
   - to perfect humans?
   - to create new, "better" life forms—plant, animal, and human?
   - to increase biomedical knowledge and procedures for treatment?

   For what other purposes do you think genetic engineering may be pursued?
2. Because, as Christians believe, we are capable of both good and evil intentions and actions and because, as history shows, applications of new knowledge have had good and bad consequences, what, if any, responsibilities do we have to guide the emerging knowledge and technologies of genetic engineering?
3. Because of the Jewish and Christian teaching that humankind has a unique status among created beings, having been created in the image of God, is it an equally important truth to maintain that our bodily substances—including blood, tissue, organs, genetic material—also have unique status?
4. Is human genetic material, because it contains the "building blocks" of human life and because alterations to germ cells could be inherited by future generations, especially not to be

manipulated? Should mixing genetic material from human and nonhuman organisms be allowable? Why? Why not?

SUGGESTIONS FOR STUDY LEADERS

1. Ask some participants to review before this session the creation stories in Genesis, the books of Job and Ecclesiastes, Psalm 8, and the first chapter of the Gospel of John. Ask them to look for values and insights from these sections in the Bible as well as any other sections they may think of that are relevant for the issues discussed in this section of the study report.

2. On newsprint in front of the group, record their findings, and then add any additional values or insights that others in the group believe reflect a particularly religious perspective on these issues.

3. In preparation for the next session, ask participants to read the articles in Appendix C by Dr. Lewis Thomas and Dr. Leon Kass.

• SECTION FOUR •

Values, Decisions, and Public Policy

In view of the theological and ethical considerations mentioned in the preceding parts of this report, we on the Panel of Bioethical Concerns believe that genetic engineering raises many complex questions of public policy. While affirming the tremendous value of understanding the genetic code and genetic-engineering techniques, we feel the many existing and anticipated consequences and by-products of our new powers must be thoughtfully explored. Early discussions of an issue may shape a debate for years to come. Very early in public debates, we determine who is to be involved and which issues are to be stressed. Elements that are disregarded now will be especially difficult to introduce at a later time.

Mindful of the ever-increasing speed with which knowledge and technology are advancing, we think that the National Council of the Churches of Christ and its member communions must study and engage in discussion of public policy issues now, before the decisions made and patterns established become even more fixed than they are already.

The following discussion outlines our basic values stemming from our biblical heritage and our faith experience, the major regulatory and human rights issues we have perceived, and some suggested positions for inclusion in a social policy statement on bioengineering expected to be presented to the Governing Board in the future. This section addresses issues related to values, risks (physical danger to persons, species, and the environment), human rights, justice, and education and public participation.

TOWARD A VALUE STANCE

We began by making our faith statement. Our values flow directly from this interaction of our theology, our traditional Christian values, and ethical concerns with the myriad of possibilities inherent in genetic modification. These values form the basis for our personal, individual judgments and for our public policy considerations as well. The following values emerge as being the most pertinent at this moment in history.

1. *The worth of human life.* Life in general and human life in particular is sacred. It cannot be violated by undue risks or misuse. Research must contribute the maximal benefit for the most possible persons. However, throughout the process it must protect the human and civil rights of all people, especially those least able to defend themselves. No life form can be viewed as a pawn in a chess game to be surrendered at will.

2. *The interdependence of life systems.* The interrelatedness and interdependence of all human and extrahuman life forms are more obvious than ever before. We care about the many life systems within our ecological system. The imbalance of one will directly affect another. We value the quality of life in each form as it contributes to the whole biosphere.

3. *The growth process.* We understand the nature of creation as dynamic and directional. Growth is the outgrowth of two countering forces, one to stabilize and one to change. When life's systems stabilize and rest without challenge, creation's potential is impeded.

4. *Knowledge and the pursuit of truth.* Participation in the creative process requires a continuing quest "to know," to unravel the mysteries of life. The process of obtaining knowledge must also have the goal of wholeness and humanness.

5. *Responsible scientific inquiry.* Science as a method of inquiry, as a discipline, and as a profession must be primarily self-governing, yet it must invite and expect public participation both informally and through formal regulation at points

that may affect the community's well-being. Community well-being is enhanced when the scientific community engages in peer review, minimizes risk through containment and other safety procedures, and undertakes experimental authentication processes that do not endanger human beings.

6. *Participation in decisions that may affect personal or community well-being.* Social institutions or political bodies must be available to ensure informed consent, the free flow of information, education, dialogue, and group consensus around the issues raised by biogenetic capabilities.

7. *Diversity.* We value the diversity inherent in the creation, the pluralism represented by sex, racial and ethnic identity, and cultural experience. Genetic experts must resist any temptation to alter, subordinate, or eliminate these uniquenesses.

8. *Distributive justice.* We value equity and nonmanipulative relationships. Beneficial biogenetic innovation should be generally available (never coercively) to all, regardless of geographic location, economic ability, or racial lines. Economic benefits from salable products derived from publicly funded research should be shared with the public. Rights to the manufacture and sale of such products should not be held in monopoly. Private companies that have developed or have a license to produce biogenetically engineered products should expect a reasonable return on their investment. We must also respond to the request that biogenetic innovations be available to Third World persons at costs relative to their income.

RISK

Since the notion of risk is basic to many of the issues raised in this report, it is appropriate to consider this subject at the outset. One important feature of a modern, technologically dominated society is the continual impetus to expand systematic inquiry into new areas. Implicit in this expansion is a constantly changing risk factor; new knowledge and new applications of knowledge, old and new, may increase or decrease dangers to society or the environment. In addition, scientific knowledge and modern technology also uncover previously unknown dangers from past applications or misapplications of knowledge. Given that risk inevitably accompanies technological advance, the following questions recur in various guises in the balance of this report: How much risk is acceptable? From what sources? Who shall decide such matters? On what basis? To what extent should the possibility of unknown danger restrict the search for knowledge to areas where risk is already defined? Is the risk greater if we fail in the laboratory or if we succeed and then, through trial and error, establish some form of a eugenics program?

When experimentation with recombinant-DNA technology began in the early 1970s, members of the scientific community expressed grave concern over possible risks to persons and the environment. Many scientists imposed on themselves a moratorium on certain types of experimentation until initial research guidelines were developed at the famous meetings held at the Asilomar Conference Center in Pacific Grove, California, in 1974. Some scientists, including environmentalists, expressed their concern that recombinant DNA could produce novel organisms which, when released into nature, might be beyond our control. Subsequently, the National Institutes of Health (NIH) assumed responsibility for coordinating the development of research guidelines, which pertained to the types of permissible experiments and the necessary kinds of physical containment facilities.

Our study and our conversations with scientists reveal that, with minor exceptions, microbiologists perceive that laboratory experience of the past few years has shown the initial fears to have been largely unfounded. This view has been reflected in the gradual reduction of the NIH guidelines. Serious consideration was given in February 1982 to their complete elimination, but, on the advice of a national panel of scientists, ethicists, lawyers, and theologians, the regulations, although further reduced, were kept in effect. As of February 1982, three types of experiments continue to require NIH review and prior approval: those that would confer drug resistance on bacteria that are not normally resistant; those that would confer the ability to make deadly toxins on bacteria that cannot ordinarily do so; and those that would deliberately release any recombinant organisms into the environment.

These guidelines, as was the case with their more stringent predecessors, apply only to federally funded research. We have been advised by representatives of industry that, even though not technically bound, the private sector generally adheres to these guidelines. Further, one of our resources from the pharmaceutical industry noted that even though pharmaceutical companies doing privately funded research are not subject to the NIH, other government agencies, such as the Federal Drug Administration, whose approval is vital to such companies' success, compel adherence to the NIH guidelines as a matter of administrative policy.

Our Christian responsibility to prevent human suffering and to practice stewardship of the environment requires us to be alert to the physical risks associated with recombinant-DNA experimentation. Even as this report is being written, an experiment with diphtheria toxin has been declared by other scientists and doctors as potentially dangerous, even deadly to all mammals on this planet. Our concern extends to any potential use of recombinant-DNA technology for biological warfare.

One way in which risks are limited is through professional self-regulation. Microbiologists, like other professionals, have a vested interest in maintaining high standards of safe and careful research and, because of their shared specialized knowledge, are in a position to evaluate one another's work with a particularly critical eye. As the first defense against potentially harmful experimentation, we encourage vigorous self-regulation by scientists engaged in this field. Churches may enhance such regulation by nurturing the scientific community's appreciation of the social and ethical aspects of their work, which extends beyond safe laboratory procedures.

At the present time, however, we sense that voluntary self-regulation is not sufficient. Reassurances of scientists notwithstanding, the field of genetic research is still new enough, and the potential for harm great enough, that we advocate some form of explicit governmental regulation of all parties engaging in genetic research. For the time being, the existing NIH guidelines should be maintained. The application of these guidelines should be extended to nongovernmentally funded research as well.

In addition to the risk of direct and present physical harm, we note that the artificial production of living organisms bears the risk of deleterious effects on the ecosystem extending over a broad reach of space and time. Agencies of the government with responsibility over the environment, such as the Environmental Protection Agency, should be charged, if they are not already, with monitoring all projects using private and public funds that may have such impact on the environment.

In the current absence of comprehensive regulation on the federal level, nine cities (such as Cambridge and Boston) and two states (New York and Maryland) have adopted their own regulations, sometimes making compliance with the NIH guidelines a requirement within their jurisdictions for all recombinant-DNA research, whether publicly or privately funded. In the absence of federal regulation, we support such local control. In the interest of predictability and consistency, however, national regulation would be preferable as the basic mode to limit the physical risks to persons and the environment resulting from the unintentional release of pernicious bacteria created by genetic engineering without imposing on researchers the burden of complying with varying local standards. However, in particular cases of extremely hazardous activities, the federal regulations must allow some degree of local self-determination.

HUMAN RIGHTS

Access to health care. We share in the high hopes and expectations expressed to us by microbiologists and medical doctors that genetically engineered substances, such as human growth hormone and insulin, and the methodologies of genetic intervention will provide new cures for disease and the means of correcting hitherto uncorrectable defects. Although genetic engineering may eventually provide an economical means of producing substances that are now expensive to obtain, many of the medical techniques and products of genetic engineering may, at least initially, be costly. We also recognize that the funds for some of the research out of which these techniques and products will develop are provided by the federal government.

Our Christian responsibility for human rights, for the interests of the poor, compels us to reaffirm in the context of this study of genetic engineering our support for equal access. This is especially true when those services and products result from publicly funded research. Accordingly, we encourage the development of mechanisms in the public and private sectors that would facilitate the equitable distribution of the benefits of the new genetic technology.

The question of what is a fair and just distribution of the world's resources and wealth enters this discussion as well. How do we justify spending vast amounts on genetic research for procedures and products that by cost or need will benefit only a few? We stop short of saying there is no justification for we acknowledge the worth of each human life. The proportions, however, must be examined.

Research subjects. The process of developing and perfecting the technology of genetic intervention and the resultant medical products raises a new concern about the possibility of improper medical experimentation. As we focus our attention on genetic experimentation we must question any possible existence of double standards that require medical products used in the United States to meet stringent tests but that discourage the initial testing of such products in the United States. Our concern for justice also impels us to reaffirm our opposition to any involuntary experimentation on any persons, recognizing that minorities, children, the elderly, the poor, the terminally ill, and the incarcerated are particularly vulnerable to such treatment.

We affirm the need for fully informed consent, based on knowledge of all pertinent health risks, by all research subjects. We also affirm movements such as the guidelines adopted by the former HEW Ethics Advisory Board in May 1974 regarding *in vitro* fertilization and fetal experimentation. We stand firmly against any form of discrimination under the guise of genetic improvement. Knowledge regarding an individual's genetic capacity cannot be used to prohibit freedom or rights. No individual should be compelled to undergo genetic treatment, in the name of "improving" the human species.

In addition to these general principles relating to the testing of new drugs and medical procedures, we note that genetic engineering raises several new issues with respect to experimentation. For instance, the manipulation of genes in human germ or sex cells, if permitted at all, should be subject to special scrutiny, because not only the individual, but also his or her descendants, may be affected. In such cases, it may be appropriate to require the approval of a guardian *ad litem* appointed to represent the interests of the future descendants.

When some of the even more provocative topics, such as the cloning of human beings and the cross-breeding of human germ cells with those of other species, have arisen in our conversations with scientists, the possibility of such experiments occurring has often been discounted on the general basis that experiments are not performed unless there is a particular need for them. This argument assumes that there would be no valid reason for attempting these procedures. Some people would have said the same for certain previous technologies. When the "state of the art" enables such experiments, it is possible that someone might attempt them.

We know of no existing laws or regulations to prohibit such experiments. We know of no goal legitimate enough to warrant such radical experimentation. There could be no margin for error. The potential accusation from an offspring against the human maker would have no possible redress. Irrespective of consent, we ethically must call for perimeters to the use of genetic technology.

JUSTICE

Influence of business. Research in genetic engineering may be amenable to profitable commercial exploitation. Significant capital investment in businesses related to genetic engineering is called "biobusiness." Although these businesses generally have not yet become profitable, optimism about their future continues to attract the business community to participate in the development of recombinant-DNA technology. Academia and the government are the traditional partners in this area.

Thus it is fair to anticipate that a significant proportion of the beneficial products to be derived from genetic research will be made available in part by the willingness of investors to risk capital in the hope of making a profit. While many individual investors participate, the greatest infusions of capital are coming from large corporations, particularly pharmaceutical companies.

This growing commercialism, although it enables certain lines of research to progress much faster than they otherwise would, causes concern. Substantial outlays of money imply the power to influence the direction and emphasis of biogenetic study and experimentation. Some people worry that many scientists will focus on the development of commercially useful products in order to please their business partners, to the detriment of important but not immediately profitable basic research.

It is too early to determine whether such fears are warranted. If it turns out that they are, one solution may be government intervention, such as the development of tax incentives, encouraging private investment in basic genetic research. Another option is to allow greater amounts of public financing for basic research.

Patents. The United States Supreme Court ruled in 1980 that, in effect, living organisms are eligible for patent protection (*Diamond vs. Chakrabarty*). This development in law merits careful attention. At one end of the spectrum, the patenting of certain bacteria may well be no more or less significant from a Christian viewpoint than the patenting of inanimate drugs or industrial compounds. The patent system for these inanimate drugs and compounds confers limited monopolistic rights in exchange for publication of the patented item or process. It is generally regarded as an effective method of encouraging inventiveness. (An inventor is rewarded with seventeen years of exclusive rights to his or her invention and, by placing a detailed description of his or her invention in the public records, provides other inventors with the basis for still more advanced inventions.)

At the other end of the spectrum, however, we cannot question that the patenting of cells and tissues in a human being would constitute a gross violation of human dignity and individual rights. The difficult public policy question, which requires additional study by churches, is where to draw the line between substances appropriate to patent and substances inappropriate to patent. For example, the Plant Patent Act of 1930 is now becoming a controversial issue. Public researches at universities, botanical gardens, and agricultural research centers are expressing deep concern regarding secrecy and the reduction in the free flow of plant-germ plasm. Many independent seed companies are being purchased by large oil, chemical, and pharmaceutical multinationals, which, in effect, have purchased the basic building blocks for all seed lines of the future. We can question whether those who put the "finishing touches" on a hybrid plant have more right to own it and receive a royalty for it than any of those who have preserved and developed it in the past. When one adds to this dilemma the consideration that most plant-germ banks are located in and primarily controlled by First World countries, international animosity is aroused.

Two major influences will determine how we look at the patenting continuum: (1) the extent to which various kinds of genetically engineered organisms are deemed fundamental to human life and, therefore, not subject to private ownership; and (2) the extent to which it may be desirable to foster scientific and medical advances through economic incentives.

As evidenced already in the plant-patenting history, the quest for patents may affect the free flow of information concerning genetic research. Some scientists who have addressed us indicated that the flow of research information has already lessened in recent years as appreciation of the commercial potential has increased. Because other scientists indicated that they had not witnessed a reduction in communication, we cannot say at this time with absolute certainty that the confidentiality required before the issuance of a patent (at which time a sample or detailed description of the patented material must be available for public inspection) will have a detrimental impact. However, even the secrecy already resulting from ordinary competition among scientists is enough to warrant concern regarding the patent laws.

A third concern relative to patents is that our increasing

ability to produce new strains quickly may cause the genetic pool to be diminished, either inadvertently or intentionally. This decrease in the number of strains may result from attempts by the patent owners of certain organisms or types of seeds to discourage, to the extent of their power, the use of competing strains, thus enhancing the market share and profitability of their own products. Great volumes of new plant variations can be produced at accelerating speed by genetic-engineering techniques. It was in part this ability, coupled with the wide distribution of a particular corn seed lacking resistance to corn blight, which led to the destruction of a majority of U.S. corn crops in one year.

Our stewardship over the earth and the living things on it requires us to determine carefully whether such threats of extinction exist. To the extent that they do, remedial action should be taken. The abolition of patent rights with respect to living organisms, however, may not be the most appropriate response. For instance, a solution more closely directed to the problem might be the maintenance by the government of repositories for samples of various plant strains so as to ensure that extinction does not occur. Land-grant universities and private companies already maintain such collections.

Products of publicly funded research. Another question related to commercial exploitation of recombinant-DNA technology is whether the profits derived from a commercially successful product obtained by investors who risked their capital only after years of government-funded research ought to be shared in some fashion with the public.

Such financial participation might be difficult because it would necessitate distinguishing among the endless gradations of reliance on prior research. Furthermore, Congress passed a law in 1980 allowing universities and small businesses to keep patent rights to inventions developed with federal research funds. Congress is now considering a bill to extend the same right to larger businesses. Any policy of returning profits to the government might expand to cover other commercial development of basic scientific or engineering research, such as aerospace or general medical research. Even if we hesitate to advocate the payment of monies by businesses to the government in this context, we reaffirm the need for public access to the beneficial products and techniques resulting from genetic research.

EDUCATION AND PUBLIC PARTICIPATION

We have been reminded by those advocating active public involvement in the resolution of these issues that Thomas Jefferson once made the remark:

> I know of no safe depository of the ultimate powers of this society but the people themselves; and if we think them not enlightened enough to exercise their control with a wholesome discretion, the remedy is not to take it from them, but to inform their discretion [cited in *Science*, July 30, 1979].

The need for public approval of experiments and programs having a potentially broad impact has been characterized as the public's "informed consent," drawing an analogy from a concept long accepted in the context of the medical treatment of individuals. In order to participate effectively in dialogues with scientific and technological groups and to reach appropriate conclusions concerning the issues, the public must be educated and well informed. Such knowledge could help to replace distrust with mutual respect.

The scientists who met with our Panel generally, although not unanimously, favored a system of peer review rather than public review. The reason for such a preference stems from skepticism about the ability of members of the general public to comprehend and evaluate complex scientific data and technology. We can attest to this complexity from our own experience as nonscientists in carrying out the study requested by the Governing Board; but we can further attest that understanding is possible. Further, we note that many forums, such as hearings, task forces, advisory panels, and review boards, have been quite successful not only for information-sharing but also in involving the public in biomedical policy formation.

Churches need to contribute significantly to the public's awareness of the issues relating to bioengineering and to the public's participation in discussions related to their resolution. Local churches can assist in raising the level of public understanding concerning the physical phenomena involved. Such education should occur within the broader realm of assisting Christians, and the public at large, in developing adequate theological, ethical, and moral bases for making decisions in this area. We believe that the affirmations and interpretations set forth in Section Three of this report can be useful in such education. Educational programs should be developed by the National Council of Churches or its member communions for use in local congregations to facilitate the creative participation of individuals, both scientists and nonscientists, in the resolution of the issues at hand. As part of this educational venture, churches are compelled to cooperate wherever and whenever possible with other agencies, national and international, public or scientific, working toward the same goals.

Although this endeavor will be difficult, our ever-increasing ability to manipulate the genetic code is of such profound significance that churches must do everything within their power to permit an informed public to analyze and address these varied issues. We must also be prepared to address the many additional questions of public policy that are sure to arise in the future.

QUESTIONS TO CONSIDER

1. Because potential benefits and risks from developments in genetic engineering are great, should they be allowed to be distributed at random among citizens, or is society morally obligated to have benefits and risks distributed equitably? What obligations do nations, such as the United States, which has genetic-engineering technologies, have to other nations to distribute the benefits and risks equitably?
2. Should genetic-engineering businesses that prosper from scientific knowledge funded by federal money be obligated to share their profits? When experiments go wrong, should the costs of corrective efforts be paid by the businesses or by state and federal funds or both?
3. When medical benefits from genetic engineering become available, should they be provided to those who currently benefit from the existing system of access to health care—that is, those who have private insurance or who are covered by government health insurance programs, such as Medicare

or Medicaid, the majority of whom tend to be white and middle class? Or are we morally obligated to distribute the benefits of genetic engineering more equitably than we currently distribute other health-care services? Why?

4. Is federal regulation, as recommended in this section, of research protocol, businesses, states, academic institutions, and scientists in private and public institutions warranted or unnecessary?
5. Is the ideal of an informed citizenry participating in significant choices created by major developments, such as genetic engineering, an achievable and necessary goal (recall Thomas Jefferson's comment in the text)? Or is it a pie-in-the-sky dream that will only let loose the tyranny of an uninformed mob?
6. Where do you think churches should stand in regard to the recommendations proposed in this section? What particular values, which may or may not be unique to Christians, should influence the churches' stand in our pluralistic society, in which public policy is shaped by deliberation and democratic politics?

SUGGESTIONS FOR STUDY LEADERS

1. As you have presumably done in the three previous sessions, use the questions provided, above, for group discussion questions.

2. Allow at least fifteen minutes at the end of the session for each participant to fill out the questionnaire provided in the study report and give it to you before he or she leaves. The questionnaire (pages 71–81) may be duplicated for group use.

3. After the sessions are over, tally the questionnaires and send in a summary of your group's response to the National Council of Churches at the address provided on the questionnaire. Please identify your group fully, and provide a record of their concerns and reactions in addition to the information requested in the questionnaire. If you do not return the questionnaire, you still can use it to conclude your sessions.

• APPENDIX A •
A Letter to the President of the United States

June 20, 1980

We are rapidly moving into a new era of fundamental danger triggered by the rapid growth of genetic engineering. Albeit, there may be opportunity for doing good; the very term suggests the danger. Who shall determine how human good is best served when new life forms are being engineered? Who shall control genetic experimentation and its results which could have untold implications for human survival? Who will benefit and who will bear any adverse consequences, directly or indirectly?

These are not ordinary questions. These are moral, ethical, and religious questions. They deal with the fundamental nature of human life and the dignity and worth of the individual human being.

With the Supreme Court decision allowing patents on new forms of life—a purpose that could not have been imagined when patent laws were written—it is obvious that these laws must be reexamined. But the issue goes far beyond patents.

New life forms may have dramatic potential for improving human life, whether by curing diseases, correcting genetic deficiencies or swallowing oil slicks. They may also, however, have unforeseen ramifications, and at times the cure may be worse than the original problem. New chemicals that ultimately prove to be lethal may be tightly controlled or banned, but we may not be able to "recall" a new life form. For unlike DDT or DES—both of which were in wide use before their tragic side effects were discovered—life forms reproduce and grow on their own and thus would be infinitely harder to contain.

Control of such life forms by any individual or group poses a potential threat to all of humanity. History has shown us that there will always be those who believe it appropriate to "correct" our mental and social structures by genetic means, so as to fit their vision of humanity. This becomes more dangerous when the basic tools to do so are finally at hand. Those who would play God will be tempted as never before.

We also know from experience that it would be naive and unfair to ask private corporations to suddenly abandon the profit motive when it comes to genetic engineering. Private corporations develop and sell new products to make money, whether those products are automobiles or new forms of life. Yet when the products are new life forms, with all the risks entailed, shouldn't there be broader criteria than profit for determining their use and distribution? Given all the responsibility to God and to our fellow human beings, do we have the right to let experimentation and ownership of new life forms move ahead without public regulation?

These issues must be explored, and they must be explored now. It is not enough for the commercial, scientific or medical communities alone to examine them; they must be examined by individuals and groups who represent the broader public interest. In the long-term interest of all humanity, our government must launch a thorough examination of the entire spectrum of issues involved in genetic engineering to determine before it is too late what oversight and controls are necessary.

We believe, after careful investigation, that no government agency or committee is currently exercising adequate oversight or control, nor addressing the fundamental ethical questions in a major way. Therefore, we intend to request that [you] President Carter provide a way for representatives of a broad spectrum of our society to consider these matters and advise the government on its necessary role.

We also intend to ask the appropriate Congressional Committees to begin immediately a process of revising our patent laws looking to revisions that are necessary to deal with the new questions related to patenting life forms. In addition, we will ask our government to collaborate with other governments, with the appropriate international bodies, such as the UN, to evolve international guidelines related to genetic engineering.

Finally, we pledge our own efforts to examine the religious and ethical issues involved in genetic engineering. The religious community must and will address these fundamental questions in a more urgent and organized way.

Dr. Claire Randall, General Secretary
National Council of Churches

Rabbi Bernard Mandelbaum, General Secretary
Synagogue Council of America

Bishop Thomas Kelly, General Secretary
United States Catholic Conference

• APPENDIX B •
The Process of the Panel on Bioethical Concerns

The National Council of the Churches of Christ/USA Panel on Bioethical Concerns here describes the process through which this study report was developed:

This Panel on Bioethical Concerns has now concluded its work. But, like the graduation from a formal education program, it is in reality a commencement. We have examined this new celebrity called genetic engineering. Just as a parent looks at a little child and has, at the same time, both hopes and fears, so we in this age of technology see exciting possibilities and devastating uncertainties.

To maintain a perspective that is manageable the Panel limited its focus to a subsection of bioethics, genetic technology, and its implications for today's and tomorrow's societies. More specifically, the Panel on Bioethical Concerns of the NCCC agreed that its work would focus on (1) the origins of life, i.e, the creation and patenting of new life forms, the cloning of DNA molecules; (2) the modification of life forms as it affects the quality of life in humans, animals, and plants; and (3) the termination of life forms and species.

As a Panel on Bioethical Concerns, we recognize that the areas outlined for study do not cover all the related issues. Others have addressed the issues of negative eugenic programs, euthanasia, birth control, sociobiology, and possible sexual discrimination through genetic technology. Therefore we did not extend our time or resources to these important and related subjects.

Although the panel did not have as members scientific and medical experts, we were informed by such experts. We are representatives of our member communions and of the units within the NCCC. We visited scientists in Boston, in the study hall and in the laboratories. We talked to those who work within academia. From San Francisco and from Columbus, Ohio, we invited scientists, doctors, and industrialists of pharmaceutical companies to speak and discuss the issues. We heard from governmental and consumer groups who watch and monitor the work of scientists in this field. We heard "pros" and "cons" from those most closely related to the work.

Yet even as we talk and read and study, decisions are being made and the work of biotechnology proceeds. We cannot be silent. When we do speak out we run the risk of making an error in judgment. However, to ignore the issue and therefore say nothing is to err more grievously. We report our findings and concerns and share our worries so that others will be made aware of the issues surrounding us, issues already at work in our world.

• APPENDIX C •

Excerpts from a Variety of Perspectives

1. Some leading public policy issues are set out by the President's Commission for the Study of Ethical Problems in Medicine and Biomedical and Behavioral Research (President's Commission, Splicing Life, *U.S. Government Printing Office, Washington, DC, 1982, pp. 19–23).*

Although much remains to be learned in this field, knowledge is being acquired rapidly: in most areas of research, "new" means something that has been found within the past five years; in molecular biology, it often means something found within the past few months. Time and time again in the past ten years, the speed with which events have unfolded has taken well-informed observers by surprise, as noted in a major medical journal:

> While physicians won't be performing gene therapy on humans for some time, that time appears to be approaching more rapidly every day. The tempo of applications of new, basic technologies to clinical medicine continues to be astonishing.[1]

Indeed, prognostications thus far have frequently underestimated the pace of new knowledge.

The most predictable aspect of this technology may be its very unpredictability. The Commission shares the view of the religious leaders, scientists, and others in the media, government, and elsewhere: a continuing exploration is needed of the implications of this technology that has already reshaped the direction of scientific research and that could revolutionize many aspects of life in the modern world.

No attempt is made in this Report to resolve the myriad social and ethical issues generated by the ability to manipulate the basic material of living things. The Commission found that in many instances the issues had not been clearly and usefully articulated yet. A goal of this Report, therefore, is to stimulate thoughtful, long-term discussion—not preempt it with conclusions that would, of necessity, be premature. At this stage in the public discussion, the Commission believes there are at least four broad prerequisites to the development of effective public policy*: (1) educating the public about genetics and about the historical context of genetic manipulations; (2) clarifying the concerns underlying the simplistic slogans that are frequently used; (3) identifying the issues of concern in ways meaningful to public policy consideration; and (4) evaluating the need for oversight and analyzing the responsibilities and capabilities for it both within and outside government.

Educating the public. The United States is a country with ever-increasing dependence on technological and scientific expertise. Public participation in matters that may have substantial personal import often require a fundamental knowledge of highly specialized fields. Individuals who do not acquire such knowledge may hesitate to participate in the public debate, thinking the subject is too complicated for them and best left to the experts. Alternatively, public discussion can be misguided because people lack understanding of scientific facts and appreciation of the known limits and potentials of a new technology. The issues surrounding genetic engineering face both these problems.

Public policy on genetic engineering will need to draw heavily on the wisdom of "experts" who have earned the pub-

---

*The Commission uses the term "public policy" broadly to include not only formal laws and regulations but the many programs and policies of individuals and institutions that society decides are acceptable and not in need of direct collective intervention. Public policy is not limited to situations where the government has taken action; indeed, as the Report notes, the Commission concludes that many issues raised by genetic engineering are not proper subjects of government regulation, which is itself a public policy judgment.

lic's trust and respect. But an informed public is also an essential element of a democratic decision-making process. As emphasized in the Commission's report on screening and counseling for genetic conditions, it is important to include genetics in academic curricula—beginning in early grades.[2] Even with effective formal education on genetics, however, the rapid changes taking place in this field make continuing education essential. This Report seeks to contribute to that process not only by demonstrating the need for enlightened public discussion, but also by providing the reader with some basic background about this new technology. Such a background is important not only for examining significant implications of this technology, but also for distinguishing the issues that merit serious attention from fantastic scenarios that have no scientific basis.

The Commission also finds a second type of information related to gene splicing important for public discussions—an understanding of the context in which this new technology arises. Gene splicing is a revolutionary scientific technique that recasts past ideas and reshapes future directions. Even so, it does not necessarily follow that all its applications or objectives represent a radical departure from the past. Indeed, the question of whether this application differs in significant ways from previous interventions or capacities served as an important guidepost for much of the Commission's discussion of social and ethical concerns about genetic engineering. For example, do the partnerships emerging between industry and academia in regard to gene splicing differ from past interactions in ways that give rise to new concerns or require unique responses? Would replacing a defective gene with a normal one from another person to correct a blood disorder differ socially and ethically from current investigations in which bone marrow is transplanted from one person to another for the same purpose? The Commission attempts to bring this perspective to its discussion of the issues.

Clarifying concerns expressed in slogans. A complex and seemingly mysterious new technology with untapped potential is a ready target for simplistic slogans that try to capture vague fears. This is very much the case with genetic engineering. In Chapter Three, the Commission examines some of the slogans that have been invoked on both sides of the genetic engineering controversy and attempts to clarify and analyze the concerns they seem to reflect.

A recent public opinion poll, for example, found that the single area of research in which restraint on scientific inquiry was favored is "creation of new life forms."[3] But what is meant by this term? Is bacteria into which a human insulin gene has been inserted a "new life form" that ought not to be created? Is a new hybrid corn offensive? Or is the fear of a new life form really about partially human hybrids?

Concern is also expressed about gene splicing because it will cause human beings to "control evolution" or lead to "an alteration of the gene pool." But humanity's activities have always affected the gene pool. And why would tinkering with genes mean that evolution has been "controlled"?

On the other hand, arguments in favor of caution and control are sometimes met with claims of "academic freedom." What application does this principle have in discussing physical risk to other people? And how ought the value of the pursuit of knowledge be weighed against other values?

Identifying the public policy issues. The diversity of social and ethical issues implies the need for similarly varied responses. A third objective of this Report, therefore, is to organize these issues in a way that is useful both for general understanding and for the formulation of sound public policy. The Commission has focused on the various types of uncertainties associated with the uses of gene splicing techniques: evaluative or ethical uncertainty; conceptual uncertainty; and occurrence uncertainty.

The first type of uncertainty occurs when no societal consensus exists as to whether certain applications of gene splicing are beneficial or undesirable. Should research be conducted to generate means by which "positive" traits could be introduced into a person genetically—for example, by improving memory? Would this be regarded as a socially and ethically desirable application of the technology? Further uncertainty occurs because the determination of what constitutes a "defect" or "disease" varies over time and between cultures.

Conceptual uncertainty refers to the fundamental change in concepts that this new technology can engender. As noted earlier, the notion that genes, once conceived of as fixed, can now be manipulated and exchanged has been described as "unnerving." The significance of this for people's conception of their role in the universe and even for the meaning of being human underlies an important set of concerns.

Concerns like these have not typically arisen in public policy discussions. A limited number of implications of gene splicing, however, do echo issues raised by other technologies that have prompted generally uncontroversial public policy responses. The premarket testing of new drugs is one example. A consensus exists that certain outcomes would be beneficial, such as the development of safe, effective drugs, and others harmful, such as unsafe, ineffective drugs. The uncertainty involved is whether a particular outcome will occur. Policy can be directed specifically at promoting the desirable outcomes and minimizing the likelihood of harmful effects.

Occurrence uncertainty also applies to some issues that cannot be so readily addressed. As with many new technologies, the full range of scientific effects of gene splicing cannot now be predicted with complete certainty. And those effects will be expressed in a future that cannot be known in advance.

Decisions made about the future of this technology and its applications will need to be made with reference to the varied types of risks and uncertainties at stake in gene splicing. Chapter Three attempts to organize the issues in ways that will foster the development of effective public policy.

Evaluating the need for oversight. Having set out the types of risks posed by gene splicing, the Commission then considers the need for oversight of these issues. A variety of mechanisms, involving both the government and the private sector, are possible. One common feature unites all those that seem appropriate to the Commission: they draw on, but are not controlled by, gene splicing experts.

2. Dr. Lewis Thomas considers the question, Are there some things in science we should not pursue? He answers, No. (Excerpts from "The Hazards of Science." Reprinted, by permission of The New England Journal of Medicine, vol. 209 [September 19, 1980], pp. 324–26.)

I suppose there is one central question to be dealt with, and I am not at all sure how to deal with it although I am certain about my own answer to it. It is this: are there some kinds of information, leading to some sorts of knowledge, that human beings are really better off not having? Is there a limit to scientific inquiry not set by what is knowable but by what we ought to be knowing? Should we stop short of learning about some things, for fear of what we, or someone, will do with the knowledge? My own answer is a flat no, but I must confess that this is an intuitive response and I am neither inclined nor trained to reason my way through it.

There has been some effort, in and out of scientific quarters, to make recombinant DNA into the issue on which to settle this argument. Proponents of this line of research are accused of pure hubris, of assuming the rights of gods, of arrogance and outrage; what is more, they confess themselves to be in the business of making live hybrids, with their own hands. The mayor of Cambridge, Massachusetts, and the Attorney General of New York have both been advised to put a stop to it, forthwith.

It is not quite the same sort of argument, however, as the one about limiting knowledge, although this is surely part of it. The knowledge is already here, and the rage of the argument is about its application in technology. Should DNA for making certain useful or interesting proteins be incorporated into *Escherichia coli* plasmids, or not? Is there a risk of inserting the wrong sort of toxins, or hazardous viruses, and then having the new hybrid organisms spread beyond the laboratory? Is this a technology for creating new varieties of pathogens, and should it be stopped because of this?

If the argument is held to this level, I can see no reason why it cannot be settled, by reasonable people. We have learned a great deal about the handling of dangerous microbes in the last century, although I must say that the opponents of recombinant-DNA research tend to downgrade this huge body of information. At one time or another, agents as hazardous as those of rabies, psittacosis, plague and typhus have been dealt with by investigators in secure laboratories with only rare cases of self-infection of the investigators themselves, and none at all of epidemics. It takes some high imagining to postulate the creation of brand-new pathogens so wild and voracious as to spread from equally secure laboratories to endanger human life at large, as some of the arguers are now maintaining.

But this is precisely the trouble with the recombinant-DNA problem: it has become an emotional issue, with too many irretrievably lost tempers on both sides. It has lost the sound of a discussion of technologic safety, and begins now to sound like something else, almost like a religious controversy, and here it is moving toward the central issue: are there some things in science we should not be learning about?

There is an inevitably long list of hard questions to follow this one, beginning with the one that asks whether the mayor of Cambridge should be the one to decide, first off.

Maybe we'd be wiser, all of us, to back off before the recombinant-DNA issue becomes too large to cope with. If we're going to have a fight about it, let it be confined to the immediate issue of safety and security of the recombinants now under consideration, and let us by all means have regulations and guidelines to assure the public safety wherever these are indicated, or even suggested. But if it is possible let us stay off that question about limiting human knowledge. It is too loaded, and we'll simply not be able to cope with it. . . .

It does seem to me that in the biologic and medical sciences we are still far too ignorant to begin making judgments about what sorts of things we should be learning or not learning. To the contrary, we ought to be grateful for whatever snatches we can get hold of, and we ought to be out there on a much larger scale than today's, looking for more.

We should be very careful with that word hubris, and make sure it is not used when not warranted. There is a great danger in applying it to the search for knowledge. The application of knowledge is another matter, and there is hubris in plenty in our technology, but I do not believe that looking for new information about nature, at whatever level, can possibly be called unnatural. Indeed, if there is any single attribute of human beings, apart from language, that distinguishes them from all other creatures on earth, it is their insatiable, uncontrollable drive to learn things and then to exchange the information with others of the species. Learning is what we do, when you think about it. I cannot think of a human impulse more difficult to govern.

But I can imagine lots of reasons for trying to govern it. New information about nature is very likely, at the outset, to be upsetting to someone or other. The recombinant-DNA line of research is already upsetting, not because of the dangers now being argued about but because it is disturbing, in a fundamental way, to face the fact that the genetic machinery in control of the planet's life can be fooled around with so easily. We do not like the idea that anything so fixed and stable as a species line can be changed. The notion that genes can be taken out of one genome and inserted in another is unnerving. Classical mythology is peopled with mixed beings—part man, part animal or plant—and most of them are associated with tragic stories. Recombinant DNA is a reminder of bad dreams.

The easiest decision for society to make in matters of this kind is to appoint an agency, or a commission, or a subcommittee within an agency, to look into the problem and provide advice. And the easiest course for a committee to take, when confronted by any process that appears to be disturbing people or making them uncomfortable, is to recommend that it be stopped, at least for the time being.

I can easily imagine such a committee, composed of unimpeachable public figures, arriving at the decision that the time is not quite ripe for further exploration of the transplantation of genes, that we should put this off for a while, maybe until next century, and get on with other affairs that make us less uncomfortable. Why not do science on something more popular?

The trouble is, it would be very hard to stop once this line was begun. There are, after all, all sorts of scientific inquiry that are not much liked by one constituency or another, and we might soon find ourselves with crowded rosters, panels, standing committees, set up in Washington for the appraisal, and then the regulation, of research. Not on grounds of the possible value and usefulness of the new knowledge, mind you, but for guarding society against scientific hubris, against the kinds of knowledge we're better off without.

It would be absolutely irresistible as a way of spending time, and people would form long queues for membership. Almost anything would be fair game, certainly anything to do with genetics, anything relating to population control, or, on the other side, research on aging. Very few fields would get by.

The research areas in the greatest trouble would be those already containing a sense of bewilderment and surprise, with discernible prospects of upheaving present dogmas. I can think of several of these, two current ones in which I've been especially interested, and one from the remote past of 40 years ago.

First, the older one. Suppose this were the mid-1930's, and there were a Commission on Scientific Hubris sitting in Washington, going over a staff report on the progress of work in the laboratory of O. T. Avery in New York. Suppose, as well, that there were people on the Commission who understood what Avery was up to and believed his work. This takes an excess of imagining, since there were vanishingly few such people around in the 1930's, and also Avery didn't publish a single word until he had the entire thing settled and wrapped up 10 years later. But anyway, suppose it. Surely, someone would have pointed out that Avery's discovery of a bacterial extract that could change pneumococci from one genetic type to another, with the transformed organisms now doomed to breed true as the changed type, was nothing less than the discovery of a gene; moreover, Avery's early conviction that the stuff was DNA might turn out to be correct, and what then? To this day, the members of such a committee might well have been felicitating each other on having nipped something so dangerous in the very bud.

But it wouldn't have worked in any case, unless they had been equally prescient about bacteriophage research and had managed to flag down phage genetics before it got going a few years later. Science can be blocked, I have no doubt of that, or at least slowed down, but it takes very fast footwork.

Here is an example from today's research on the brain, which would do very well on the agenda of a Hubris Commission. It is the work now going on in several laboratories here and abroad dealing with the endorphins, a class of small polypeptides also referred to as the endogenous opiates. It is rather a surprise that someone hasn't already objected to this research, since the implications of what has already been found are considerably more explosive, and far more unsettling, than anything in the recombinant-DNA line of work. There are cells in the brain, chiefly in the limbic system, which possess at their surfaces specific receptors for morphine and heroin, but this is just a biologic accident; the real drugs, with the same properties as morphine, are the pentapeptide hormones produced by the brain itself. Perhaps they are switched on as analgesics at times of trauma or illness; perhaps they even serve for the organization and modulation of the physiologic process of dying when the time for dying comes. These things are not yet known, but such questions can now be asked. It is not even known whether an injection of such pentapeptides into a human being will produce a heroin-like reaction, but that kind of question will also be up for asking, and probably quite soon since the same peptides can be synthesized with relative ease. What should be done about this line of research—or rather, what should have been done about it two or three years ago when it was just being launched? Is this the sort of thing we are better off not knowing? I know some people who might think so. But if something prudent and sagacious had been done, turning off such investigations at an early stage, we would not have glimpsed the possible clue to the mechanism of catatonic schizophrenia, which was published just this month from two of the laboratories working on endorphins.

It is hard to predict how science is going to turn out, and if it is really good science it is impossible to predict. This is in the nature of the enterprise. If the things to be found are actually new, they are by definition unknown in advance, and there is no way of pretelling in advance where a really new line of inquiry will lead. You cannot make choices in this matter, selecting things you think you're going to like and shutting off the lines that make for discomfort. You either have science, or you don't, and if you have it you are obliged to accept the surprising and disturbing pieces of information, even the overwhelming and upheaving ones, along with the neat and promptly useful bits. It is like that.

And even if it were possible to call most of the shots in advance, so that we could make broad selections of the general categories of new knowledge that we like, leaving out the ones we don't have a taste for, there would always be slips, leaks, small items of shattering information somehow making their way through. . . .

The only solid piece of scientific truth about which I feel totally confident is that we are profoundly ignorant about nature. Indeed, I regard this as the major discovery of the past 100 years of biology. It is, in its way, an illuminating piece of news. It would have amazed the brightest minds of the 18th-century enlightenment to be told by any of us how little we know, and how bewildering seems the way ahead. It is this sudden confrontation with the depth and scope of ignorance that represents the most noteworthy contribution of 20th-century science to the human intellect.

We are, at last, facing up to it. In earlier times, we either pretended to understand how things worked or ignored the problem, or simply made up stories to fill the gaps. Now that we have begun exploring in earnest, doing serious science, we are getting glimpses of how huge the questions are, and how far from being answered. Because of this, these are hard times for the human mind, and it is no wonder that we are depressed. It is not so bad being ignorant if you are totally ignorant; the hard thing is knowing in some detail the reality of ignorance, the worst spots and here and there the not-so-bad spots, but no true light at the end of any tunnel nor even any tunnels that can yet be trusted. Hard times, indeed.

But we are making a beginning, and there ought to be some satisfaction, even exhilaration, in that. The method works. There are probably no questions we can think up that can't be answered, sooner or later, including even the matter of consciousness. To be sure, there may well be questions we can't think up, ever, and therefore limits to the reach of human intellect that we will never know about, but that is another matter. Within our limits, we should be able to work our way through to all our answers, if we keep at it long enough, and pay attention.

I am putting it this way, with all the presumption and confidence that I can summon, to raise another, last question. Is this hubris? Is there something fundamentally unnatural, or intrinsically wrong, or hazardous for the species, in the ambition that drives us all to reach a comprehensive understanding of nature, including ourselves? I cannot believe it. It would seem to me a more unnatural thing, and more of an offense against nature, for us to come on the same scene endowed as we are with curiosity, filled to overbrimming as we are with questions, and naturally talented as we are for the asking of clear questions, and then for us to do nothing about it, or worse, to try to suppress the questions. This is the greater danger for our species, to try to pretend that we are another kind of animal, that we do not need to satisfy our curiosity, that we can get along somehow without inquiry and exploration, and experimentation, and that the human mind can rise above its ignorance by simply asserting that there are things it has no need to know. This, to my way of thinking, is the real hubris, and it carries danger for us all.

3. Dr. Leon Kass argues that we must use caution and perhaps not pursue certain genetic engineering technologies. (Excerpts from "The New Biology: What Price Relieving Man's Estate?" Science, vol. 174, no. 4011 [November 19, 1971], pp. 779–88.) Copyright 1971 by AAAS. Used by permission.

The advances we shall examine are fruits of a large, humane project dedicated to the conquest of disease and the relief of human suffering. The biologist and physician, regardless of their private motives, are seen, with justification, to be the well-wishers and benefactors of mankind. Thus, in a time in which technological advance is more carefully scrutinized and increasingly criticized, biomedical developments are still viewed by most people as benefits largely without qualification. The price we pay for these developments is thus more likely to go unrecognized. For this reason, I shall consider only the dangers and costs of biomedical advance. As the benefits are well known, there is no need to dwell upon them here. My discussion is deliberately partial. . . .

Genetic engineering, when fully developed, will wield two powers not shared by ordinary medical practice. Medicine treats existing individuals and seeks to correct deviations from a norm of health. Genetic engineering, in contrast, will be able to make changes that can be transmitted to succeeding generations and will be able to create new capacities, and hence to establish new norms of health and fitness.

Nevertheless, one of the major interests in genetic manipulation is strictly medical: to develop treatments for individuals with inherited diseases. Genetic disease is prevalent and increasing, thanks partly to medical advances that enable those affected to survive and perpetuate their mutant genes. The hope is that normal copies of the appropriate gene, obtained biologically or synthesized chemically, can be introduced into defective individuals to correct their deficiencies. This *therapeutic* use of genetic technology appears to be far in the future. Moreover, there is some doubt that it will ever be practical, since the same end could be more easily achieved by transplanting cells or organs that could compensate for the missing or defective gene product.

Far less remote are technologies that could serve *eugenic* ends. Their development has been endorsed by those concerned about a general deterioration of the human gene pool and by others who believe that even an undeteriorated human gene pool needs upgrading. Artificial insemination with selected donors, the eugenic proposal of Herman Muller,[4] has been possible for several years because of the perfection of methods for long-term storage of human spermatozoa. The successful maturation of human oocytes in the laboratory and their subsequent fertilization now make it possible to select donors of ova as well. But a far more suitable technique for eugenic purposes will soon be upon us—namely, nuclear transplantation, or cloning. Bypassing the lottery of sexual recombination, nuclear transplantation permits the asexual reproduction or copying of an already developed individual. The nucleus of a mature but unfertilized egg is replaced by a nucleus obtained from a specialized cell of an adult organism or embryo (for example, a cell from the intestines or the skin). The egg with its transplanted nucleus develops as if it had been fertilized and, barring complications, will give rise to a normal adult organism. Since almost all the hereditary material (DNA) of a cell is contained within its nucleus, the renucleated egg and the individual into which it develops are genetically identical to the adult organism that was the source of the donor nucleus. Cloning could be used to produce sets of unlimited numbers of genetically identical individuals, each set derived from a single parent. Cloning has been successful in amphibians and is now being tried in mice; its extension to man merely requires the solution of certain technical problems.

Production of man-animal chimeras by the introduction of selected nonhuman material into developing human embryos is also expected. Fusion of human and nonhuman cells in tissue culture has already been achieved. . . .

Basic Ethical and Social Problems in the Use of Biomedical Technology

First, we must recognize that questions of use of science and technology are always moral and political questions, never simply technical ones. All private or public decisions to develop or to use biomedical technology—and decisions *not* to do so—inevitably contain judgments about value. This is true even if the values guiding those decisions are not articulated or made clear, as indeed they often are not. Secondly, the value judgments cannot be derived from biomedical science. This is true even if scientists themselves make the decisions.

These important points are often overlooked for at least three reasons.

1) They are obscured by those who like to speak of "the control of nature by science." It is men who control, not that abstraction "science." Science may provide the means, but men choose the ends; the choice of ends comes from beyond science.

2) Introduction of new technologies often appears to be the result of no decision whatsoever, or of the culmination of decisions too small or unconscious to be recognized as such. What can be done is done. However, someone is deciding on the basis of some notions of desirability, no matter how self-serving or altruistic.

3) Desires to gain or keep money and power no doubt influence much of what happens, but these desires can also be formulated as reasons and then discussed and debated.

Insofar as our society has tried to deliberate about questions of use, how has it done so? Pragmatists that we are, we prefer a utilitarian calculus: we weigh "benefits" against "risks," and we weigh them for both the individual and "society." We often ignore the fact that the very definitions of "a benefit" and "a risk" are themselves based upon judgments about value. . . .

What Is to Be Done?

First, we sorely need to recover some humility in the face of our awesome powers. The arguments I have presented should make apparent the folly of arrogance, of the presumption that we are wise enough to remake ourselves. Because we lack wisdom, caution is our urgent need. Or to put it another way, in the absence of that "ultimate wisdom," we can be wise enough to know that we are not wise enough. When we lack sufficient wisdom to do, wisdom consists in not doing. Caution, restraint, delay, abstention are what this second-best (and, perhaps, only) wisdom dictates with respect to the technology for human engineering.

If we can recognize that biomedical advances carry significant social costs, we may be willing to adopt a less permissive, more critical stance toward new developments. We need to reexamine our prejudice not only that all biomedical innovation is progress, but also that it is inevitable. Precedent certainly favors the view that what can be done will be done, but is this necessarily so? Ought we not to be suspicious when technologists speak of coming developments as automatic, not subject to human control? Is there not something contradictory in the notion that we have the power to control all the untoward consequences of a technology, but lack the power to determine whether it should be developed in the first place?

What will be the likely consequences of the perpetuation of our permissive and fatalistic attitude toward human engineering? How will the large decisions be made? Technocratically and self-servingly, if our experience with previous technologies is any guide. Under conditions of laissez-faire, most technologists will pursue techniques, and most private industries will pursue profits. We are fortunate that, apart from the drug manufacturers, there are at present in the biomedical area few large industries that influence public policy. Once these appear, the voice of "the public interest" will have to shout very loudly to be heard above their whisperings in the halls of Congress. These reflections point to the need for institutional controls.

Scientists understandably balk at the notion of the regulation of science and technology. Censorship is ugly and often based upon ignorant fear; bureaucratic regulation is often stupid and inefficient. Yet there is something disingenuous about a scientist who professes concern about the social consequences of science, but who responds to every suggestion of regulation with one or both of the following: "No restrictions on scientific research," and "Technological progress should not be curtailed." Surely, to suggest that *certain* technologies ought to be regulated or forestalled is not to call for the halt of *all* technological progress (and says nothing at all about basic research). Each development should be considered on its own merits. Although the dangers of regulation cannot be dismissed, who, for example, would still object to efforts to obtain an effective, complete, global prohibition on the development, testing, and use of biological and nuclear weapons?

The proponents of laissez-faire ignore two fundamental points. They ignore the fact that not to regulate is as much a policy decision as the opposite, and that it merely postpones the time of regulation. Controls will eventually be called for—as they are now being demanded to end environmental pollution. If attempts are not made early to detect and diminish the social costs of biomedical advances by intelligent institutional regulation, the society is likely to react later with more sweeping, immoderate, and throttling controls.

The proponents of laissez-faire also ignore the fact that much of technology is already regulated. The federal government is already deep in research and development (for example, space, electronics, and weapons) and is the principal sponsor of biomedical research. One may well question the wisdom of the direction given, but one would be wrong in arguing that technology cannot survive social control. Clearly,the question is not control versus no control, but rather what kind of control, when, by whom, and for what purpose.

Means for achieving international regulation and control need to be devised. Biomedical technology can be no nation's monopoly. The need for international agreements and supervision can readily be understood if we consider the likely American response to the successful asexual reproduction of 10,000 Mao Tse-tungs.

To repeat, the basic short-term need is caution. Practically, this means that we should shift the burden of proof to the *proponents* of a new biomedical technology. Concepts of "risk" and "cost" need to be broadened to include some of the social and ethical consequences discussed earlier. The probable or possible harmful effects of the widespread use of a new technique should be anticipated and introduced as "costs" to be weighed in deciding about the *first* use. The regulatory institutions should be encouraged to exercise restraint and to formulate the grounds for saying "no." We must all get used to the idea that biomedical technology makes possible many things we should never do.

But caution is not enough. Nor are clever institutional arrangements. Institutions can be little better than the people who make them work. However worthy our intentions, we are deficient in understanding. In the *long* run, our hope can only lie in education: in a public educated about the meanings and limits of science and enlightened in its use of technology; in scientists better educated to understand the relationships between science and technology on the one hand, and ethics and politics on the other; in human beings who are as wise in the latter as they are clever in the former.

Notes

SECTION TWO: FUNDAMENTAL FACTS

1. For a history of developments in biochemical and molecular genetics, see Horace Freeland Judson, *The Eighth Day of Creation* (New York: Simon & Schuster, 1979).

2. "Thus, the underlying issue in the recombinant DNA research debate is the accommodation of knowledge-thrust and the public interest. Shall unfolding knowledge determine our desired future or shall our hoped-for future contribute to choices regarding the direction of knowledge-thrust?"—Clifford Grobstein, "Regulation and Basic Research: Implications of Recombinant DNA," *S. Cal. L. Rev.*, 51:1181, 1199 (1978).

3. Raoul E. Beneviste and George J. Todaro, "Gene Transfer Between Eukaryotes," *Science*, 217:1202 (1982).

4. In higher organisms that reproduce sexually, a high degree of genetic variation is produced by the normal process of crossing-over of genes in the germ cells. Crossing-over, like the other processes, involves the formation of new combinations of genes.

5. Gerald M. Rubin and Allan C. Spalding, "Genetic Transformation of Drosophilia Germ Line Chromosomes," *Science*, 218:348 (1982).

6. Victor A. McKusick, *Mendelian Inheritance in Man*, 6th ed. (Baltimore: Johns Hopkins University Press, 1982).

7. Richard Roblin, "Human Genetic Therapy: Outlook and Apprehensions," in George K. Chacko, ed., *Health Handbook* (New York: Elsevier-North Holland Pub. Co., 1979), pp. 103, 108–12.

8. Jean L. Marx, "Still More About Gene Transfer," *Science*, 218:459 (1982).

9. Thomas E. Wagner et al., "Microinjection of a Rabbit B-globin Gene into Zygotes and Its Subsequent Expression in Adult Mice and Their Offspring," *Proc. National Academy of Science*, 78:6376 (1981).

10. Ralph L. Brinster et al., "Somatic Expression of Herpes Thymidine Kinase in Mice Following Injection of a Fusion Gene into Eggs," *Cell*, 27:223 (1981).

11. Richard D. Palmiter et al., "Differential Regulation of Metallothionein—Thymidine Kinase Fusion Genes in Transgenic Mice and Their Offspring," *Cell*, 29:701 (1982).

12. Bob Williamson, "Reintroduction of Genetically Transformed Bone Marrow Cells into Mice," *Nature*, 284:397 (1980).

13. Council of Europe Parliamentary Assembly, 23d Ordinary Session, *Recommendation 934*, Strasbourg (1982).

14. Nicholas Wade, *The Ultimate Experiment* (New York: Walker & Company, 1977), p. 2.

15. As Chief Justice Burger observed, some of the arguments presented against issuance of a patent for the oil-eating bacteria "remind us that, at times, human ingenuity seems unable to control fully the forces it creates—that with Hamlet, it is sometimes better 'to bear those ills we have than fly to others that we know not of.'" Diamond v. Chakrabarty, 447 U.S. 303, 316 (1980).

APPENDIX C: Excerpts from a Variety of Perspectives

1. Lawrence D. Grouse, "Restriction Enzymes, Interferon, and the Therapy for Advanced Cancer," *Journal of the American Medical Association*, 247:1742 (1982).

2. Screening and Counseling for Genetic Conditions (Washington, DC: U.S. Government Printing Office, 1983), note 3 at section 3, chap. 2.

3. John Walsh, "Public Attitude Toward Science Is Yes, But—," *Science*, 215:270 (1982).

4. H. J. Muller, *Science*, 134:643 (1961).

Questionnaire

Your comments on this study report are requested. Please fill out the questionnaire on the following pages. You are encouraged to add comments to accompany the questionnaire when you return it. You may care to add a summary of your answers to the discussion questions provided at the end of each section or perhaps elaborate on a particular point about which you have some expertise or strong feelings.

If you use this questionnaire, please send the questionnaire and comments by August 1, 1984, to:

Panel on Bioethical Concerns
National Council of Churches of Christ/USA
Room 880
475 Riverside Drive
New York, NY 10115

1. The public policy recommendations in this report are derived from eight "values stemming from our biblical heritage and our faith experience," described fully in the beginning of section four. Next to each summary statement, mark a "1" if you strongly agree that it is an important and valid biblical and theological value, a "2" if the biblical, theological assumptions are open for questioning, or a "3" if you strongly disagree that the value statement is an important and valid biblical and theological value:

   ____ Life in general and human life in particular is sacred. It cannot be violated by undue risks or misuse. Research must contribute the maximal benefit for the most possible persons. However, throughout the process it must protect the human

FOLD AND TEAR

and civil rights of all people, especially those least able to defend themselves.

____ The interrelatedness and interdependence of all human and extrahuman life forms are more obvious than ever before. We value the quality of life in each form as it contributes to the whole biosphere.

____ We understand the nature of creation as dynamic and directional. Growth is the outgrowth of two countering forces, one to stabilize and one to change.

____ Participation in the creative process requires a continuing quest "to know," to unravel the mysteries of life. The process of obtaining knowledge must also have the goal of wholeness and humanness.

____ Science as a method of inquiry, as a discipline, and as a profession must be primarily self-governing, yet it must invite and expect public participation both informally and through formal regulation at points that may affect the community's well-being.

____ Social institutions or political bodies must be available to ensure informed consent, the free flow of information, education, dialogue, and group consensus around the issues raised by biogenetic capabilities.

____ We value the diversity inherent in the creation, the pluralism represented by sex, racial and ethnic identity, and cultural experience. Genetic experts must resist any temptation to alter, subordinate, or eliminate these uniquenesses.

____ Beneficial biogenetic innovation should be generally available (never coercively) to all, regardless of geographic location, economic ability, or race.

Describe below additional biblical or theological

FOLD AND TEAR

points you believe confirm or deny the eight values identified in this study report:

____

____

____

____

____

2. At the conclusion of the fourth section of this study report, fourteen public policy recommendations are made. Next to each summary statement provided below, mark a "1" if you strongly agree with the recommendation, a "2" if you do not feel strongly about the recommendation, or a "3" if you strongly disagree:

____ We encourage vigorous self-regulation by scientists engaged in this field.

____ We advocate some form of explicit governmental regulation of all parties engaged in genetic research. For the time being, the existing NIH guidelines should be maintained. The application of guidelines should be extended to nongovernmentally funded research as well.

____ Agencies of the government with responsibility over the environment, such as the Environmental Protection Agency, should be charged, if they are not already, with monitoring all projects using private and public funds that may have an impact on the environment.

____ In the absence of federal regulation, we support local control [municipal and state laws that require compliance with NIH guidelines for all genetic research within their jurisdictions whether the work receives public or private funding]. However, national regulation would be preferable as the basic mode to limit the physical risks to persons and the environment resulting

FOLD AND TEAR

from the unintentional release of pernicious bacteria created by genetic engineering without imposing on researchers the burden of complying with varying local standards.

____ We encourage the development of mechanisms in the public and private sectors that would facilitate the equitable distribution of the benefits of the new genetic technology.

____ How do we justify spending vast amounts on genetic research for procedures and products which by cost or need will benefit only a few? We stop short of saying there is no justification for we acknowledge the worth of each human life. The proportions, however, must be examined.

____ We must question any possible existence of double standards that require medical products used in the United States to meet stringent tests but that discourage the initial testing of such products in the United States. Our concern for justice also impels us to reaffirm our opposition to any involuntary experimentation on any persons, recognizing that minorities, children, the elderly, the poor, the terminally ill, and the incarcerated are particularly vulnerable to such treatment.

____ No individual should be compelled to undergo genetic treatment, in the name of "improving" the human species.

____ The manipulation of genes in human germ or sex cells, if permitted at all, should be subject to special scrutiny, because not only the individual, but also his or her descendants, may be affected.

____ We know of no existing laws or regulations to prohibit such experiments [cloning of human beings, cross-breeding of human germ cells with those of other species]. We call for perimeters to the use of genetic technology.

____ It is fair to anticipate that a significant proportion

FOLD AND TEAR

of the beneficial products to be derived from genetic research will be made in part by the willingness of investors to risk capital in the hope of making a profit. It is too early to determine whether such fears are warranted. If it turns out that they are, one solution may be government intervention, such as the development of tax incentives encouraging private investment in basic genetic research. Another option is to allow greater amounts of public financing for basic research.

____ In regard to patents, new genetic products and technologies, profits, and public access, the difficult public policy question is where to draw the line between substances appropriate to patent and substances inappropriate to patent. We can question whether those who put the "finishing touches" on a hybrid plant have more right to own it and receive a royalty for it than any of those who have preserved and developed it in the past. As evidenced already in the plant-patenting history, the quest for patents may affect the free flow of information concerning genetic research. The secrecy already resulting from ordinary competition among scientists is enough to warrant concern regarding the patent laws.

____ We affirm the need for public access to the beneficial products and techniques resulting from genetic research.

____ In order to participate effectively in dialogues with scientific and technological groups and to reach appropriate conclusions concerning the issues, the public must be educated and well informed. Churches need to contribute significantly to the public's participation in discussions related to their resolution.

FOLD AND TEAR

3. What is your general reaction to this study report:
   A. ☐ Very helpful ☐ Helpful ☐ Somewhat helpful ☐ Not helpful
   B. ☐ Learned a great deal ☐ Learned a little ☐ Learned very little
   C. ☐ Issues clearly stated ☐ Some issues clear, others not ☐ Very confusing
   D. ☐ Clearly written ☐ Adequately written ☐ Poorly written
   E. The best part was ______________________.
      The weakest part was ______________________.

4. Information about respondent(s).
   A. If you are filling out this questionnaire as a summary of your *personal* response, please fill out the following information:

Age
☐ 25 or under
☐ 26–35
☐ 36–45
☐ 46–55
☐ 56–65
☐ 66 or older

Religious identity
name of denomination
____
☐ no religious identity
☐ clergy ☐ laity

Gender
☐ male ☐ female

Educational background (mark highest level achieved)
☐ below high school
☐ high school
☐ college degree
☐ graduate degree, master's
☐ graduate degree, doctorate; field ______

Occupation ____________

(Because this information is anonymous, it can be used only for statistical purposes.)

   B. If this questionnaire represents a summary of a *group's* response, please fully identify the group (for example: Committee on Bioethcs, Synod of the Trinity; Sunday Adult Forum, St. Paul's A.M.E. Church, Atlanta, GA).

____

FOLD AND TEAR

FOLD AND TEAR